Analytical Methods for use in Geochemical Exploration

R.E. Stanton, D.I.C., F.I.M.M., A.R.A.C.I.
Chief Chemist, Hunting Technical Services Ltd.,
Borehamwood, Herts.

A Halsted Press Book

John Wiley & Sons
New York

Published in the U.S.A.
by Halsted Press, a Division
of John Wiley & Sons, Inc.
New York

Library of Congress Cataloging in Publication Data

Stanton, Ronald Ernest.
Analytical methods for use in geochemical exploration

"A Halsted Press book."
Bibliography: p.
1. Trace elements—Analysis. 2. Geochemistry.
Analytic. I. Title.
QE516.T85S7 551.9'028 76-26050
ISBN 0-470-98920-3

Printed in Great Britain by Unwin Brothers Ltd,
London and Woking.

Contents

Preface

This work succeeds a previous volume (Stanton, 1966), which deals exclusively with colorimetric methods of analysis. Since that time analysis by atomic-absorption spectrophotometry has become commonplace, and it is now the leading analytical technique in use for geochemical exploration. It is both sensitive and rapid, although requiring considerable capital outlay.

Some methods of analysis by X-ray fluorescence spectrometry are also described, and the application of emission spectrography is discussed in a chapter contributed by Dr C. H. James, to whom I am most grateful. Both these techniques require expensive equipment.

A chapter on cold extraction techniques is included at the request of many geologist friends.

I would like to express my appreciation of the assistance given by Mrs A. J. Hardwick and Mrs G. Bellevue de Sylva in the preparation of the manuscript.

St. Albans
1976

R.E.S.

1. Statistical Control of Analysis

For geochemical exploration it is essential to know the precision of analysis from the low background level through threshold to the low anomalous range. A 'statistical series' of samples as suggested by Craven (1953-4) is suitable as a set of control samples. These samples must be numbered at random, and it is advisable, although not always practicable, that the analyst should be unaware as to which are control samples and which are routine. The control samples should be included at intervals of approximately one in ten.

The results obtained on the control samples are subjected to a statistical treatment, that described by James (1970) being preferred to Craven's original approach. It should be noted that the method is a check on precision and not accuracy, although it could be adapted for this latter purpose.

Procedure

1. Obtain two samples of at least 250g each of sieved material, one (L) having a low background content and the other (H) having a content two or three times the threshold level of the element to be determined.
2. Homogenize each sample.
3. Weigh nine parts of L and mix thoroughly with one part of H.
4. Similarly, prepare further samples from L and H, using the ratios 8:2, 7:3, 6:4, 4:6, 3:7, 2:8 and 1:9 of L and H, respectively.
5. Increase this series to ten samples by using the individual low and high components.
6. Applying James' statistical treatment, x = the proportion of H in each sample and has the values 0, 0.1, 0.2, 0.3, 0.4, 0.6, 0.7, 0.8, 0.9 and 1.0.
7. Multiply the analytical result (y) of each sample by the value of x for that sample, and sum the products to give Σxy.
8. Obtain Σy by summation of all the analytical results.
9. Solve the following equations for a and b:
   $1.1a = \Sigma xy - \Sigma y/2$ and $b = \Sigma y/10 - a/2$
10. Calculate $\hat{H}$ (the high component) which equals $a + b$, where b is $\hat{L}$ (the low component).
11. Obtain a calculated value (A) for each sample: $A = \hat{H}x + \hat{L}(1 - x)$.
12. Obtain the difference ($y - A$) for each sample. The accuracy of

the calculation may be checked at this stage by summing the differences, since $\Sigma(y-A)$ should equal zero. The calculated values are usually rounded off to the nearest whole number, and these approximations may prevent a summation to zero but by no more than ±5.

13. For each sample obtain the square of the difference and sum these squares to give $\Sigma(y-A)^2$.

14. Calculate the variance $\Sigma(y-A)^2/8$.

15. Take the square root of the variance to obtain the standard deviation (σ).

16. The mean precision at the 95% confidence level is given as a percentage by

$$\frac{2 \times \sigma \times 100}{(\hat{H}+\hat{L})/2}$$

2. Reagents

Alamine 336. $(CH_3(CH_2)_6CO)_3N$. Tri-caprylyl tertiary amine.
This is a colourless liquid, immiscible with water but miscible with organic solvents such as toluene. Contact with the skin must be avoided as it may cause severe irritation.

Arsenazo III

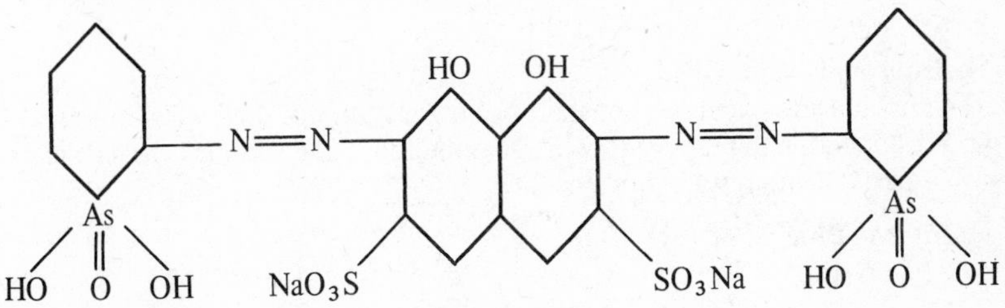

Di-sodium salt of 2,7–di(*o*-arsonophenylazo)-1,8-hydroxynaphthalene-3,6–disulphonic acid.

This compound reacts with many elements, including uranium and thorium. The thorium complex has a green colour.

Bromine, Br_2
Bromine is a dark, reddish brown, fuming liquid, the vapour of which irritates all parts of the respiratory system. It is particularly severe on the eyes and nose, and must always be used in a fume cupboard. It is only slightly soluble in water, but is very soluble in hydrobromic acid. Its boiling point is 58.8°C and it freezes at -7.9°C.

iso–Butylmethylketone. $(CH_3)_2CHCH_2COCH_3$, 4–Methylpentan–2–one, iso-Propyl acetone, Hexone.
This is a colourless liquid with a boiling point of 116°C and a flash point of 17°C. It is a narcotic, and irritates the eyes and nose. As an extracting medium immiscible with water, it is used extensively to separate and concentrate many elements for analysis by atomic-absorption spectrophotometry.

Chloroacetic Acid. $CH_2Cl\,COOH$.
This is a colourless, crystalline compound, soluble in water. It is very corrosive, and attacks the skin to cause severe blistering.

1,2–Dichloroethane. $CH_2Cl\,CH_2Cl$.
The boiling point of this solvent is 83.5°C and it has a flash point of 13°C. It is a colourless liquid, immiscible with and more dense than water, which disolves rubber but not silicone rubber. The vapour irritates the eyes and nose and it is a powerful narcotic, causing damage to the liver and kidney. If the reagent gives a high blank determination for boron, it may be purified by passing it through a column of chromatographic grade alumina, Brockman activity 1 or 2.

Formic Acid. $H\,COOH$.
In concentrations of 90 to 100% this is a colourless, fuming liquid with an irritant vapour; it attacks the skin giving rise to blisters. The 98 to 100% reagent has a flash point greater than 77°C.

Hydrobromic Acid. HBr.
The concentrated acid is a constant-boiling mixture of 48% w/w, boiling at 126°C, and is slightly yellow-brown in colour. The vapour is irritating to all parts of the respiratory system.

Methylene Blue

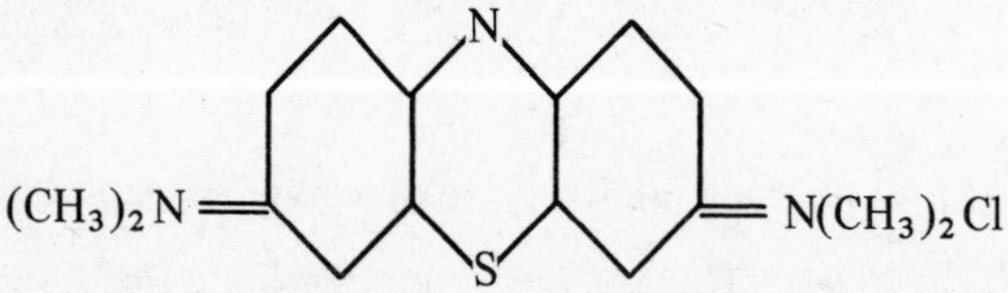

$(CH_3)_2N=$ [phenothiazine ring: N, S] $=N(CH_3)_2Cl$

This reagent is usually obtained as the di-hydrate. It is a redox indicator and a microscopical stain. A blue complex is formed with the fluoroborate ion.

Petroleum spirit
There are many solvents under this heading and they are classified by the boiling range, the product boiling between 80° and 100°C being particularly useful. They are mixtures of aliphatic hydrocarbons obtained by fractional distillation of crude petroleum. In general, they are acute narcotics, and chronic poisoning may give rise to nervous and digestive disturbances.

Sodium Bisulphate. $NaHSO_4$. Sodium hydrogen sulphate.
This compound is very similar to potassium bisulphate, which it replaces in the determination of vanadium (p. 21); it may be used as a substitute for potassium bisulphate in many other determinations.

3. Methods of Colorimetric Analysis

Bismuth

Procedure

1. Weigh 0.5g of sample into a borosilicate test-tube (19 x 150mm).
2. Mix with 1.5g of potassium bisulphate and fuse until a quiescent melt is obtained.
3. Leach with 5ml of 4M nitric acid on a sand-tray or in a boiling water-bath.
4. Add 5ml of water, mix and leave to settle.
5. Pipette 5ml of the clear solution into a separating funnel (50ml).
6. Add 10ml of EDTA solution and mix.
7. Add ammonia solution dropwise until the reddish-brown colour persists
8. Add 4M nitric acid dropwise until the reddish-brown colour just disappears.
9. Add 1ml of ammonia solution and mix.
10. Add 5ml of potassium cyanide solution and mix.
11. Add 1ml of sodium diethyldithiocarbamate solution and mix.
12. Add 2.5ml of chloroform and shake the stoppered funnel for 1 minute.
13. Transfer the organic phase to a test-tube (16 x 150mm) and cork the tube.
14. Compare with a standard series.
15. Bismuth in p.p.m. = 4 x μg of matching standard.

Standards

1. To five separating funnels (50ml) add, respectively, 0, 10, 20, 40 and 80 μg of bismuth.
2. Dilute to 5ml with 2M nitric acid.
3. Add 2ml of iron solution.
4. Continue as described in Stages 6–13 of the Procedure.

Reagents

Potassium bisulphate. Fused, powder.
Concentrated nitric acid. Sp.gr. 1.42, analytical reagent grade.
4M nitric acid. Mix 250ml of the concentrated acid with 750ml of water.

2M nitric acid. Mix 125ml of the concentrated acid with 875ml of water.
Ammonium ferric sulphate. Analytical reagent grade.
Sulphuric acid. Sp.gr. 1.84, analytical reagent grade.
Iron solution. Dissolve 8.64g of ammonium ferric sulphate in 100ml of water and 10ml of sulphuric acid, then dilute to 1 litre with water. This solution will contain 1mg of iron per ml.
EDTA. Ethylene diaminetetra-acetic acid, di-sodium salt.
EDTA solution. Dissolve 50g of EDTA in 1 litre of water.
Ammonia solution. Sp.gr. 0.91, analytical reagent grade.
Potassium cyanide. Analytical reagent grade.
Potassium cyanide solution. Dissolve 25g of the solid in 500 ml of water
Sodium diethyldithiocarbamate solution. Dissolve 1 g in 100ml of water. Prepare freshly as required.
Chloroform. Analytical reagent grade.
Bismuth nitrate. Penta-hydrate, analytical reagent grade.
Standard bismuth solutions. 100 μg of bismuth per ml: dissolve 116mg of bismuth nitrate in 250ml of 4M nitric acid and dilute to 500ml with water.
20 μg bismuth per ml: dilute 20ml of the 100 μg per ml solution to 100ml with 2M nitric acid.

The sample is decomposed by fusion with potassium bisulphate, most silicates being attacked to leave a residue of silica, while the other major components form sulphates. The colour of the iron(III)–EDTA complex is utilized to adjust the pH between 5 and 6 at Stages 7 and 8 of the Procedure. The formation of the diethyldithiocarbamates of metals such as iron(III), mercury(II), silver and copper(II) is prevented by use of an ammoniacal solution in the presence of cyanide and EDTA, the optimum pH range being 10-11. The method as described covers the range 20-160 p.p.m. of bismuth.

The use of cyanide necessitates vigorous precautions, the following procedure being recommended. All sinks must have water flowing through them, and no acid or cyanide solutions must be discharged into them. Beakers containing sodium hydroxide solution must be placed at each sink to receive waste acid solutions, and discarded test solutions should be kept separately. The method of disposal of the waste solutions should be settled by consultation with the appropriate local authority. The analyst must rinse his hands under running water immediately prior to pipetting.

An antidote must be readily available at all times, and should be administered rapidly if cyanide poisoning is suspected. The following recipe is recommended (Royal Institute of Chemistry, 1961):

Solution A: dissolve 158g of ferrous sulphate crystals (B.P.) and 3g of citric acid crystals in 1 litre of water.

Solution B: dissolve 60g of anhydrous sodium carbonate in 1 litre of water.

Place 50ml of A in a 6oz wide-necked bottle closed by a polythene covered cork and labelled clearly CYANIDE ANTIDOTE A; similarly for CYANIDE ANTIDOTE B. Both bottles should bear the legend 'Mix the whole contents of bottles A and B and swallow the mixture'.

The method of analysis is based on the work of Stanton (1971b).

Boron

Procedure

1. Weigh 0.1g of sample into a small polythene beaker.
2. Add 2.5ml of 5M sulphuric acid.
3. Add 0.5ml of hydrofluoric acid.
4. Stir with a polythene rod and cover with polythene.
5. Leave to stand at room temperature for 2 hours.
6. Add 2ml of water and mix.
7. Leave to stand for 15 minutes.
8. Transfer 1ml of the clear solution into a test-tube calibrated at 15ml.
9. Add 1ml of methylene blue solution.
10. Dilute to 15ml with water.
11. Add 5ml of 1,2-dichloroethane.
12. Stopper the tube with a silicone rubber bung and shake it vigorously for 30 seconds.
13. Allow the phases to separate and compare with a standard series.
14. Boron in p.p.m. = 50 x μg of matching standard.

Standards

1. To each of eleven polythene beakers add, respectively, 0,1.0, 2.0, 3.0, 4.0, and 5.0 μg of boron using the standard solution containing 2 μg of boron per ml, and 7.5, 10.0, 15.0, 20.0 and 25.0 μg of boron using the standard solution containing 10 μg of boron per ml.
2. Dilute to 2.5ml with 5M sulphuric acid.
3. Continue as described in Stages 3-12 of the Procedure.

Reagents

Sulphuric acid. Sp.gr.1.84, analytical reagent grade.
5M sulphuric acid. Cautiously add 270ml of the concentrated acid to 500ml of water with mixing, and dilute to 1 litre with water.
Hydrofluoric acid. 40% w/w, analytical reagent grade.
Methylene blue solution. Dissolve 80 mg in 100ml of water.
1, 2-dichloroethane.
Sodium tetraborate. Deca-hydrate, analytical reagent grade.
Standard boron solutions. 100 μg of boron per ml: dissolve 441 mg of sodium tetraborate in 5M sulphuric acid and dilute with this acid to 500ml.

10 μg of boron per ml: dilute 10ml of the 100 μg per ml solution to 100ml with 5M sulphuric acid.

2 μg of boron per ml: dilute 2ml of the 100 μg per ml solution to 100ml with 5m sulphuric acid.

Borosilicate glassware must not be used in this determination, and even soda glass contains sufficient boron to give appreciable contamination. It is convenient to use polythene ice-cube trays for the acid treatment of the samples, a second tray being used as a cover. Pipettes should be made of quartz glass or polythene, and test-tubes may be of either quartz or lead glass. Bark corks are unsatisfactory as they absorb a considerable amount of methylene blue which may contaminate subsequent tests.

This method was developed for geochemical studies in which the boron was derived from the mineral colemanite, and was readily attacked in the cold by a mixture of dilute hydrofluoric and sulphuric acids to form fluoroborate. Consequently, application of this method to other materials may be restricted by this method of sample decomposition. The acid concentrations are not critical, but must be kept at the same constant level for both samples and standards, and 5M sulphuric acid is used so that alkaline samples produce no significant variation in acidity. The volume of hydrofluoric acid is kept low to minimize the colour extracted from the zero standard; the level of the blank determination may be reduced by purification of the 1,2-dichloroethane (p. 4). The sample weight may be increased to 250mg without any other alteration to the procedure.

Fluoroborate ions react with methylene blue to form a blue complex extractable by 1, 2-dichloroethane. The slight colour extracted from the zero standard is dependent upon the final aqueous phase acidity, decreasing with decreasing acidity while the efficiency of extraction of the fluoroborate complex increases. It is therefore important that the final acid concentration should be constant, and if more than 1ml of sample solution is used for analysis, the preparation of the standard series must be adjusted to obtain similar conditions. Likewise, when less than 1ml is used, compensating amounts of hydrofluoric and sulphuric acids must be added to the aliquot. The standard series shows an increasing intensity of blue from a slightly blue zero, and it is stable for two days.

There is interference from dichromate and nitrate, but these ions are unlikely to be present. Arsenate and mercury also have an adverse effect.

The method covers a range of 5–250 p.p.m. of boron, and the work may be organised so that 150 samples may be analysed per man day. It is the work of Stanton and McDonald (1966).

Molybdenum

Procedure

1. Weigh 0.25g of sample into a borosilicate test-tube (16 x 150mm).
2. Mix with 1g of potassium bisulphate and fuse until a quiescent melt is obtained.
3. Leach with 5ml of 6M hydrochloric acid on a sand tray or in a boiling water-bath.
4. Add 5ml of 6M hydrochloric acid, mix and leave to settle.
5. Pipette a 5ml aliquot of the clear solution into a test-tube. (16 x 150mm) calibrated at 2, 7 and 9ml and containing 2ml of reducing solution.
6. Mix and leave to stand for 2 minutes.
7. Add 2ml of potassium iodide solution and mix.
8. Add 1ml of dithiol solution and mix.
9. Leave to stand for 2 minutes.
10. Add 0.5ml of petroleum spirit.
11. Stopper the tube with a rubber bung and shake it vigorously for 90s.
12. Compare with a standard series.
13. If the intensity of colour in the solvent phase exceeds that of the highest standard, repeat from Stage 5 using an aliquot of 0.5ml or less, adding 2ml of iron solution and diluting to the 7ml mark with 6M hydrochloric acid.
14. Molybdenum in p.p.m. =

$$\frac{40 \times \mu g \text{ of matching standard}}{\text{ml of aliquot}}$$

Standards

1. To thirteen test-tubes (16 x 150mm) calibrated at 2, 7 and 9ml. and containing 2ml of reducing solution, add respectively 0, 0.2, 0.4, 0.6, 0.8, 1.0 and 1.5 μg of molybdenum using a standard solution of molydenum containing 1 μg of molybdenum per ml. and 2.0, 3.0, 4.0, 5.0, 7.0 and 10.0 μg of molybdenum using a solution containing 10 μg of molybdenum per ml.
2. Add 2ml of iron solution.
3. Dilute to the 7ml mark with 6M hydrochloric acid.
4. Continue as described in Stages 6-11 of the Procedure.

Reagents

Potassium bisulphate. Fused, powder.
Hydrochloric acid. Sp.gr. 1.18, analytical reagent grade.
6M hydrochloric acid. Mix 240ml of the concentrated acid with 200ml of water or use the constant-boiling mixture.

Citric acid. Mono-hydrate, analytical reagent grade.
Ascorbic acid. Analytical reagent grade.
Reducing solution. Dissolve 75g of citric acid and 150g of ascorbic acid in water and dilute to 1 litre.
Potassium iodide. Analytical reagent grade.
Potassium iodide solution. Dissolve 200g of potassium iodide in 200ml of water.
Zinc dithiol.
Thioglycollic acid. Sp. gr. 1.33.
Ethyl alcohol. Absolute.
Sodium hydroxide. Pellets, analytical reagent grade.
Dithiol solution. Add 2ml of ethyl alchol to 0.3g of zinc dithiol, followed by 4ml of water, 2g of sodium hydroxide and 1ml of thioglycollic acid. Mix well until the sodium hydroxide has dissolved and dilute to 50ml with water. Mix with 50ml of potassium iodide solution and store in a refrigerator when not in use.
Petroleum spirit. Boiling range 80–100°C. Analytical reagent grade.
Ammonium ferric sulphate. Analytical reagent grade.
Iron solution. Disolve 5g of ammonium ferric sulphate crystals in 500ml of 6M hydrochloric acid.
Sodium molybdate. Di-hydrate, analytical reagent grade.
Standard molybdenum solutions. 100 μg of molybdenum per ml: dissolve 126mg of sodium molybdate in 6M hydrochloric acid and dilute to 500ml with this acid.

10 μg of molybdenum per ml: dilute 10ml of the 100 μg per ml solution to 100ml with 6M hydrochloric acid.

1 μg of molybdenum per ml: dilute 10ml of the 10 μg per ml solution to 100ml with 6M hydrochloric acid.

Molybdenum reacts with dithiol over the range 0.5 to 5M hydrochloric acid. The greater acidity is preferred since the tolerance towards copper increases with increasing acidity, especially in the presence of iodide. Interference from iron is prevented by reduction with ascorbic acid to the ferrous state, but since ferrous ions cause a slight loss of intensity of the molybdenum-dithiol complex, the addition of iron to the standard series is necessary. A minimum period of two minutes is required for the reduction of ferric ions.

The standard series ranges from a colourless zero through an increasing intensity of green colour in the solvent phase. Two minutes are required for complete formation of the molybdenum complex before solvent extraction; this period must not be exceeded since the tungsten complex is likely to be formed on prolonged standing, although this reaction is inhibited by citric acid.

The inclusion of potassium iodide in the dithiol solution serves to increase the density of this solution, thus promoting rapid mixing when it is added to the sample solution. In its absence, when dithiol solution is added slowly enough to remain at the top of the sample solution, a heavy, grey precipitate is formed which inhibits the formation of the molybdenum-dithiol complex and obscures the colour of the subsequent solvent phase. The presence of thioglycollic acid in the dithiol solution serves to retard oxidation of this reagent.

When using an aliquot of 5ml, the maximum limits of interfering elements are (approximately) for copper (grey precipitate at the interface) 8.0%, for tungsten (blue solvent phase) 1.6%, for arsenic (yellow solvent phase) 0.16% and for selenium (red precipitate at the interface) 160 p.p.m. Tin forms a red complex but when present in the sample as cassiterite it is not taken into solution following fusion with potassium bisulphate.

The method covers the range 1–80 p.p.m. using an aliquot of 5ml. which may be extended to 400 p.p.m. if an aliquot of 0.1ml is taken. A productivity of 100 samples per man-day is easily achieved.

This procedure was developed by Stanton and Hardwick (1967). A modification by Stanton (1970a) makes use of a perchloric acid solution of the sample. For organic-rich samples, the method of Stanton, Mockler and Newton (1973) is recommended. This requires digestion with a mixture of nitric and perchloric acids, evaporation to dryness, and dissolution of the residue in dilute hydrochloric acid; the method then proceeds as described above. Other methods are given by North (1956), B.R.G.G.M. (1957), Ward, Lakin, Canney et al. (1963), Marshall (1964) and Baker (1965).

Palladium and Platinum

Procedure

1. Weigh 10g of sample into a silica crucible.
2. Heat at 600°C for 30 minutes.
3. Transfer the residue to a beaker (400ml).
4. Add 50ml of bromine solution.
5. Leave to stand overnight then boil off the excess of bromine.
6. Dilute to about 100ml with water.
7. Filter through a Millipore filter and wash with 2M hydrobromic acid.
8. Return the filtrate and washings to the clean beaker.
9. Add an excess of stannous chloride solution; 25–50ml will be required.
10. Heat to boiling, add 1ml of tellurium solution and continue to boil for 10 minutes.
11. Leave to cool, then filter through a Millipore filter and wash with 2M hydrobromic acid.
12. Dissolve the precipitate off the filter paper into a beaker (150ml) using about 10ml of hot aqua regia.

13. Evaporate the solution to dryness without baking.
14. Add 5ml of concentrated hydrochloric acid and warm to dissolve salts.
15. Dilute to 50ml with water in a calibrated flask.
16. Pipette an aliquot of 25ml into a beaker (150ml).
17. Add 0.5ml of gold solution.
18. Add 1ml of phosphomolybdic acid solution.
19. Add 5ml of formic acid.
20. Add 10ml of sodium hydroxide solution.
21. Add anti-bumping granules and cover the beaker with a watch-glass.
22. Boil the solution for 15 minutes.
23. Add 5ml of formic acid and boil for 5 minutes.
24. Add another 5ml of formic acid and again boil for 5 minutes.
25. Cool to room temperature and compare with a standard series.
26. If the intensity of colour is greater than that of the highest standard, repeat from Stage 16 of the Procedure using a smaller aliquot diluted to 25ml with M hydrochloric acid.
27. Palladium + platinum in p.p.m. =

$$\frac{5 \times \mu g \text{ of matching standard}}{\text{ml of aliquot}}$$

Standards

1. To six beakers (150ml) add, respectively, 0, 0.2, 0.4, 0.6, 0.8, and 1.0μg of platinum.
2. Dilute to 25ml with M hydrochloric acid.
3. Continue as described in Stages 17-24 of the Procedure. Prepare the standard series simultaneously with the samples.

Reagents

Hydrobromic acid. Sp. gr. 1.49, analytical reagent grade.

2M hydrobromic acid. Mix 226ml of the concentrated acid with 774ml of water.

Bromine. Analytical reagent grade.

Bromine solution. Mix 20ml of bromine with 1 litre of concentrated hydrobromic acid.

Hydrochloric acid. Sp gr. 1.18, analytical reagent grade.

6M hydrochloric acid. Mix 440ml of the concentrated acid with 400ml of water.

M hydrochloric acid. Mix 80ml of the concentrated acid with 800ml of water.

Stannous chloride. Di-hydrate, analytical reagent grade.

Stannous chloride solution. Dissolve 500g of stannous chloride in concentrated hydrochloric acid and dilute to 1 litre with this acid.

Tellurium dioxide.
Tellurium solution. Dissolve 625mg of tellurium dioxide in 1 litre of 6M hydrochloric acid.
Nitric acid. Sp. gr. 1.42, analytical reagent grade.
Aqua regia, Mix 400ml of concentrated hydrochloric acid with 100ml of concentrated nitric acid. Prepare as required.
Gold solution. Dissolve 100mg of gold in bromine solution, boil to expel the excess of bromine and dilute to 1 litre with concentrated hydrobromic acid.
Orthophosphoric acid. Sp. gr. 1.75, analytical reagent grade.
Ammonium molybdate. Tetra-hydrate, analytical reagent grade.
Phosphomolybdic acid solution. Dissolve 10g of ammonium molybdate in 100ml of water and add 4ml of orthophosphoric acid slowly while stirring.
Formic acid. Sp. gr. 1.20, analytical reagent grade.
Sodium hydroxide. Pellets, analytical reagent grade.
Sodium hydroxide solution. Dissolve 350g of the pellets in water and dilute to 1 litre.
Millipore filter. 47mm diam., 0.45μ, HAWP 00.
Standard platinum solutions. 100μg of platinum per ml: dissolve 50mg of platinum in aqua regia and evaporate to dryness without baking. Dissolve the residue in 6M hydrochloric acid and dilute to 500ml with this acid.

1μg of platinum per ml: dilute 1ml of the 100μg per ml solution to 100ml with M hydrochloric acid. Prepare freshly as required.

This method depends on the reduction of phosphomolybdic acid by formic acid to give molybdenum blue, the reaction being catalysed by palladium and platinum. Mercury behaves similarly, but it is volatilized during the preliminary ignition of the sample. Gold also enhances the reaction in the presence, but not in the absence, of palladium and platinum; however, gold in excess of 50μg produces no further increase in the colour. Rhodium and iridium interfere to some extent, but they are only partially attacked by a mixture of hydrobromic acid and bromine, and their occurrence in appreciable concentrations is unlikely. Stages 2, 4, 5, 12, 13 and 22 to 24 must all be carried out in a fume cupboard.

At Stage 5, it is important to note the presence of bromine vapour after overnight standing. If there is no excess of bromine, which may be the case with sulphide-rich samples, the palladium and platinum will not necessarily be completely in solution, and further bromine must be added. Occasional agitation of the beaker and its contents will assist the dissolution of palladium and platinum during the standing period.

Stannous chloride is used to reduce tellurium to the elemental state, when the precious metals are co-precipated. After filtering, this precipitate is washed with dilute hydrobromic acid to prevent the hydrolysis of tin. The excess of stannous chloride may be judged by the reduction of ferric ions, which are usually present at a sufficiently high concentration for this reaction to be obvious.

The colorimetric stage of the Procedure is very difficult to achieve and may necessitate considerable practice. It is essential that the hotplate heats uniformly over its surface, and it should be adjusted to a setting that will enable the solutions to boil continuously for the appropriate time, but without the salts coming out of solution. A strip of asbestos should be placed across the front of the hotplate to shield the beakers from the cold air coming into the fume cupboard. The actual times stated may be varied to suit local conditions, but the temperature must be the same for samples as for standards.

It is not possible to distinguish between palladium and platinum and a combined result is obtained. When equal amounts of the two metals are treated they produce identical intensities of colour. They are reduced to the colloidal state and may adhere to the beaker when the test solution is discarded. It is advisable, therefore, to rinse the beaker with warm bromine solution after a preliminary wash with tap water.

The standard series increases in intensity of blue colour from a light blue zero. The range covered is 0.02 - 0.2 p.p.m. using an aliquot of 25ml, which may be extended to 1 p.p.m. by use of a 5ml aliquot. If the result is expected to be less than 0.1 p.p.m., the whole 10g sample may be used.

This method is based on that of Stanton (1975).

Thorium

Procedure

1. Weigh 0.5g of sample into a small teflon beaker.
2. Add 5ml of nitric acid.
3. Add 5ml of hydrofluoric acid.
4. Heat gently for 15 minutes.
5. Add 5ml of 10M hydrochloric acid.
6. Evaporate to dryness.
7. Add 10ml of 5M sulphuric acid.
8. Evaporate until the liquid sulphuric acid has disappeared.
9. Dissolve the residue in 10ml of 6M hydrochloric acid.
10. Transfer to a test-tube (16 x 150mm) and leave to settle.
11. Pipette an aliquot of 5ml. into a separating funnel (100ml).
12. Add 15ml of 10M hydrochloric acid.
13. Add 20ml of alamine solution.
14. Stopper the funnel and shake it for 2 minutes.
15. Allow the phases to separate and transfer the aqueous phase to a test-tube (19 x 150mm).
16. Pipette an aliquot of 10ml of the aqueous phase into a test-tube (16 x 150mm).

17. Add 1ml of acid mixture and mix.
18. Add 1ml of dilute arsenazo III solution and mix.
19. Leave for 10 minutes then compare with a standard series.
20. If the colour of the solution is more green than that of the highest standard, repeat as described in Stages 16-19 of the Procedure but using an aliquot of 1ml diluted to 10ml with 9M hydrochloric acid.
21. Thorium in p.p.m. =

$$\frac{80 \times \mu g \text{ of matching standard}}{\text{ml of aliquot at Stage 16}}$$

Standards

1. To twenty-one test-tubes (16 x 150mm) calibrated at 10ml, add, respectively, 0, 1, 2, 3, 4, 5, 6, 7, 8, 9, 10, 11, 12, 13, 14, 15, 16, 17, 18, 19, and 20μg of thorium.
2. Dilute to 10ml with 9M hydrochloric acid.
3. Continue as described in Stages 17 - 19 of the Procedure.

Reagents

Sulphuric acid. Sp. gr. 1.84, analytical reagent grade.
5M sulphuric acid. Cautiously add 270ml of the concentrated acid to 500ml of water with mixing, and dilute to 1 litre with water.
Nitric acid. Sp. gr. 1.42, analytical reagent grade.
Hydrofluoric acid. 40% w/w, analytical reagent grade.
10M hydrochloric acid. Sp. gr. 1.16, analytical reagent grade.
9M hydrochloric acid. Dilute 900ml of the 10M acid to 1 litre with water.
6M hydrochloric acid. Dilute 600ml of the 10M acid to 1 litre with water.
Toluene. Sulphur-free grade.
Alamine 336. Tri-caprylyl tertiary amine.
Alamine solution. Dilute 100ml of alamine 336 to 1 litre with toluene.
Ascorbic acid. Analytical reagent grade.
Oxalic acid. Di-hydrate, analytical reagent grade.
Acid mixture. Dissolve 10g of ascorbic acid and 80g of oxalic acid in water and dilute to 1 litre. Store in a refrigerator when not in use.
Arsenazo III.
Sodium hydroxide. Pellets, analytical reagent grade.
Sodium hydroxide solution. Dissolve 5g of the pellets in 1 litre of water.
Arsenazo III stock solution. Dissolve 0.25g of arsenazo III in 10ml of sodium hydroxide solution and dilute to 100ml with water.
Dilute arsenazo III solution. Dilute 20ml of the stock solution to 200ml with water. Prepare freshly each day.
Thorium nitrate. Hexa-hydrate, analytical reagent grade.
Standard thorium solutions. 100μg of thorium per ml: dissolve 127mg of thorium nitrate in 2ml of 10M hydrochloric acid and evaporate to

dryness. Dissolve the residue in 9M hydrochloric acid and dilute to 500ml with this acid.

10μg of thorium per ml: dilute 10ml of the 100μg per ml solution to 100ml with 9M hydrochloric acid.

The sample is decomposed by heating with nitric, hydrofluoric and hydrochloric acids; silicates are attacked and volatilized by hydrofluoric acid, and free silica is also eliminated. The three acids are removed by heating with sulphuric acid, the excess of this acid being evaporated and a solution of the sample residue obtained in 6M hydrochloric acid. This solution is adjusted to 9M with hydrochloric acid and most elements are removed by extraction with alamine 336. Thorium remains in the aqueous phase together with zirconium, titanium and the rare earth elements. Zirconium is masked by oxalic acid, but titanium and the rare earth elements interfere; up to 0.8% of titanium and 0.06% of cerium can be tolerated. The tolerance towards the other rare earth metals improves with increase in atomic number.

Thorium forms a green-coloured complex with arsenazo III while the zero standard is magenta in colour. As the thorium concentration increases through the standard series, a blue component is gradually introduced into the magenta colour, and at 20μg the colour is blue to blue-grey. Further increases in thorium concentration results in colour changes through blue-green to the green colour of the thorium complex with no excess of arsenazo III; no doubt the standard series could be extended beyond 20μg if desired. The standard series is stable for 24 hours if kept in the dark when not in use.

The range 2-80p.p.m. is covered as described, being extended to 800 p.p.m. by the use of 1ml at Stage 16 of the Procedure. The work may be organized so that an average productivity of 100 samples per man-day is achieved.

This procedure is described by Stanton (1971a).

Tin

Procedure

1. Weigh 1g of sample into a borosilicate test-tube (18 x 180mm).
2. Mix thoroughly with 1g of ammonium iodide.
3. Heat, with frequent rotation and agitation of the tube, until all the ammonium iodide has sublimed.
4. Discard the loose residue by inverting the tube.
5. Allow to cool and add 5ml of M hydrochloric acid.
6. Leach on a sand-tray or in a boiling water-bath and allow to settle.
7. Pipette a 1ml aliquot of the clear solution into a test-tube (18 x 180mm) calibrated at 5, 10, 15 and 20ml.
8. Dilute to 5ml with buffer solution.

9. Mix, and leave to stand for 10 minutes.
10. Compare with a standard series.
11. If the test solution is more pink than the highest standard, add successive 5ml volumes of buffer solution until the colour is grey, mix well and leave for 10 minutes before comparing with the standard series.
12. If the test solution is still more pink than the highest standard, repeat from Stage 7 of the Procedure using an aliquot of 0.1ml and also add 0.5ml of M hydrochloric acid.
13. Tin in p.p.m. =

$$\frac{\mu g \text{ of matching standard} \times \text{ml of final solution}}{\text{ml of aliquot}}$$

Standards

1. To eleven test-tubes (18 x 180mm.) calibrated at 5ml, add, respectively, 0, 0.25, 0.5, 1.0, 1.5, 2.0, 2.5, 3.0, 3.5, 4.0 and 5.0μg of tin.
2. To each of the first six tubes add 0.5ml of M hydrochloric acid.
3. Add 0.2ml of gelatine solution.
4. Dilute to 5ml with buffer solution.
5. Mix, and allow to stand for 10 minutes before using.

Reagents

Ammonium iodide. Fine crystals.
Hydrochloric acid. Sp. gr. 1.18, analytical reagent grade.
M hydrochloric acid. Mix 40ml of the concentrated acid with 400ml of water.
Sodium hydroxide. Pellets, analytical reagent grade.
Chloroacetic acid. Analytical reagent grade.
Ascorbic acid.
Gallein.
Methylene blue.
Ethyl alcohol. Absolute.
Reagent solution. Dissolve 0.1g of gallein in 100ml of ethyl alcohol by warming gently, then filter through a Whatman No. 1 filter paper (11cm diam.). Dissolve 0.015g of methylene blue in 100ml of water. Combine these two solutions in equal proportions.
Buffer solution. Dissolve 26g of sodium hydroxide in 400ml of water, and when cold, mix this solution with a cold solution of 106g of chloroacetic acid and 5g of ascorbic acid in 400ml of water. Dilute to 1 litre with water and mix. This solution should have a pH value of 2.55±0.1. Mix this solution with 10ml of reagent solution.
Gelatine solution. Dissolve 0.5g of gelatine in 100ml of water by heating gently.
Standard tin solutions. 100μg of tin per ml: dissolve 500mg of tin powder

in 50ml of concentrated hydrochloric acid and dilute to 500ml with water.

5μg of tin per ml : dilute 5ml of the 100μg per ml solution to 100ml with M hydrochloric acid.

By heating the sample with ammonium iodide, tin present as cassiterite is converted into stannic iodide, which, together with the excess of ammonium iodide, iodine, and some other metal iodides, sublimes on the upper, cooler parts of the tube. With iron-rich samples there will be a vigorous evolution of iodine. Heating is continued until all the ammonium iodide has volatilized completely and the residue has the appearance of a liquid and is glowing red. The non-volatile residue is discarded and the sublimate is dissolved in dilute hydrochloric acid. Stannic iodide is readily hydrolysed with precipitation of the hydroxide if kept in solution at an acidity less than M hydrochloric acid. For this reason, both the sample leach and the standard solutions must be in this concentration of acid.

The decomposition of the sample and the leaching stage are facilitated if the concentration of tin in the sample is such that a sample weight of 0.2g and a leach volume of 10ml may be used.

If there is need to retain the sample solution overnight it must be filtered first. Although the bulk of the loose residue has been discarded to make it easier to obtain a clear sample solution at Stage 7, some residue inevitably remains. Tin is lost from solution when left in contact with the residual solid matter, possibly as a result of either absorption by carbon or flocculation of hydroxychlorostannate complexes by colloids from the soil.

The chloracetate buffer solution maintains the pH value of the final solution within the range 2.0 to 2.5, and is prepared from chloroacetic acid and sodium hydroxide rather than from the acid and its sodium salt. Sodium chloroacetate cannot always be obtained sufficiently pure; it is often contaminated by sodium carbonate which results in the pH value being much too high. However, care must be exercised to ensure that the sodium hydroxide used is not wet or otherwise unsatisfactory, and it is advisable to check the pH value of the buffer solution using a pH meter if this is available. In preparing the buffer solution, heat must be kept to a minimum, since it can cause degradation of the chloroacetic acid and result in a permanent yellow colour after reduction of the iodine at Stage 8 of the Procedure.

Although the optimum range of pH for the reaction of tin with gallein is 2.0 to 3.0, the narrower limits of 2.0 to 2.5 must be maintained since the reduction of ferric ions is incomplete above pH 2.5. Below pH 2.0 the reaction of tin with gallein is not quantitative.

Ascorbic acid is present in the buffer solution to reduce the iodine liberated during the decomposition of the sample. The iodide thus produced, together with that from the excess of ammonium iodide, plays an essential part in the method by preventing gallein from reacting with copper, iron, lead and manganese.

Antimony, molybdenum and tungsten also react with gallein, their complexes being similar in colour to that of the tin compound. A titanium complex is also formed which is yellow in low concentrations.

The tin complex is a lake, and on standing for several hours it will give a pink deposit on the bottom of the test-tube, leaving a solution coloured only by the excess of gallein and the methylene blue. To prevent this from happening in the standard series, gelatine is included as a protective colloid. The standard series ranges from green at zero through grey to purple and then pink. Methylene blue functions as a screening agent, and more or less may be used to suit the individual taste. It may even be omitted altogether, and water substituted in preparing the reagent solution. In this event, the standard series ranges from a yellow zero through orange to increasing intensities of pink. The concentration range is 0.5 - 100 p.p.m., which may be extended to 1000 p.p.m. by using an aliquot of 0.1ml, and eighty samples may be analysed per man-day.

This procedure is based on that of Stanton and McDonald (1961-2). An alternative method is described by Marranzino and Ward (1958).

Tungsten

Procedure

1. Weigh 0.25g of sample into a borosilicate test-tube (16 x 150mm).
2. Add 1g of potassium bisulphate, mix, and fuse until a quiescent melt is obtained.
3. Leach with 5ml of concentrated hydrochloric acid on a sand-tray or in a boiling water-bath.
4. Mix with a further 5ml of concentrated hydrochloric acid and leave to settle.
5. Pipette a 5ml aliquot of the clear solution into a test-tube (16 x 150mm) calibrated at 5 and 10ml and containing 5ml of stannous chloride solution.
6. Mix, and heat in a boiling water-bath for 10 minutes.
7. Add 1ml of dithiol solution to the warm test solution, mix and leave to stand for 10 minutes.
8. Add 0.5ml of iso-amyl acetate.
9. Stopper the tube with a rubber bung and shake it vigorously for 30s.
10. Compare with a standard series.
11. If the intensity of colour in the solvent phase exceeds that of the highest standard, repeat from Stage 5 of the Procedure using a smaller aliquot and diluting to the 10ml mark with concentrated hydrochloric acid.

12. Tungsten in p.p.m. =

$$\frac{40 \times \mu g \text{ of matching standard}}{\text{ml of aliquot}}$$

Standards

1. To eleven test-tubes (16 x 150mm) calibrated at 5 and 10ml, add, respectively, 0, 0.5, 1.0, 2.0, 3.0, 4.0, 5.0, 6.0, 7.0, and 10μg of tungsten.
2. Dilute to 5ml with concentrated hydrochloric acid.
3. Add 5ml of stannous chloride solution.
4. Continue as described in Stages 6-9 of the Procedure.

Reagents

Potassium bisulphate. Fused, powder.
Hydrochloric acid. Sp. gr. 1.18, analytical reagent grade.
Stannous chloride. Di-hydrate, analytical reagent grade.
Stannous chloride solution. Dissolve 100g of stannous chloride in concentrated hydrochloric acid and dilute to 1 litre with this acid.
Zinc dithiol.
Thioglycollic acid.
Ethyl alcohol. Absolute.
Sodium hydroxide. Pellets, analytical reagent grade.
Dithiol solution. Add 2ml of ethyl alcohol to 0.3g of zinc dithiol followed by 4ml of water, 2g of sodium hydroxide and 1ml of thioglycollic acid. Mix well until the sodium hydroxide has dissolved and dilute to 100ml with water. Store in a refrigerator when not in use.
Iso-amyl acetate. Analytical reagent grade.
Sodium tungstate. Di-hydrate, analytical reagent grade.
Standard tungsten solutions. 100μg of tungsten per ml : dissolve 90mg of sodium tungstate in concentrated hydrochloric acid and dilute to 500ml with this acid.

5μg of tungsten per ml : dilute 5ml of the 100μg per ml solution to 100ml with concentrated hydrochloric acid.

After decomposition of the sample by fusion with potassium bisulphate, the sample solution is reduced with stannous chloride and the tungsten-dithiol complex is formed in the warm solution. A period of 10 minutes is required for the complex to develop fully, and it also allows the solution to cool before extraction with iso-amyl acetate. The standard series ranges from a colourless zero through an increasing intensity of blue colour in the solvent phase, and it is stable for 24 hours. When a sample shows a greater intensity of colour than the highest standard, it may be diluted with iso-amyl acetate to obtain an approximate value as an aid to the choice of an optimum smaller aliquot.

Interference occurs from copper concentrations greater than 4000 p.p.m., molybdenum greater than 8000 p.p.m., and mercury greater than 400 p.p.m.

When it is necessary to determine both molybdenum and tungsten on the same sample, the tungsten procedure should be followed using an aliquot of 4ml. Another aliquot of 4ml may be taken for molybdenum, diluted with 4ml of water and analysed as described by Stanton (1970a)

The range covered is 2-80 p.p.m., which may be extended to 4000 p.p.m. by using an aliquot of 0.1ml. A productivity of 100 samples per man-day is easily achieved.

This method is described by Stanton (1970b). Other methods are described by Ward (1951), North (1956), B.R.G.G.M. (1957) and Bowden (1964).

Vanadium

Procedure

1. Weigh 0.5g of sample into a borosilicate test-tube (19 x 150mm).
2. Mix with 1.5g of sodium bisulphate and fuse until a quiescent melt is obtained.
3. Leach with 5ml of 4M nitric acid on a sand-tray or in a boiling water-bath.
4. Add 5ml of water, mix and leave to settle.
5. Pipette 5ml of the clear solution into a test-tube (18 x 180mm). calibrated at 5 and 10ml.
6. Add 1ml of concentrated nitric acid and boil for 5s.
7. Add 2ml of phosphotungstate solution.
8. Dilute to 10ml with water and mix.
9. Heat in a boiling water-bath for 10 minutes.
10. Leave to cool for 30 minutes.
11. Add 2ml of iso-amyl alcohol.
12. Stopper the tube with a rubber bung and shake it vigorously for 15s.
13. Compare with a standard series.
14. If the intensity of yellow colour in the solvent phase exceeds that of the highest standard, dilute with iso-amyl alcohol, mixing by gently agitating the tube, until the colour can be matched within the standard range.
15. If the intensity of colour is still too great, repeat from Stage 5 of the Procedure using a smaller aliquot diluted to 5ml with 2M nitric acid.
16. Vanadium in p.p.m. =

$$\frac{10 \times \mu g \text{ of matching standard} \times \text{ml of solvent phase}}{\text{ml of aliquot}}$$

Standards

1. To twelve test-tubes (18 x 180mm) calibrated at 5 and 10ml, add, respectively, 0, 2, 4, 6, 8, 10, 15, 20, 25, 40 and 50μg of vanadium.
2. Dilute to 5ml with 2M nitric acid.
3. Continue as described in Stages 6 – 12 of the Procedure.

Reagents

Sodium bisulphate. Anhydrous; if necessary, crush to pass 80 mesh.
Nitric acid. Sp. gr. 1.42, analytical reagent grade.
4M nitric acid. Dilute 250ml of the concentrated acid to 1 litre with water.
2M nitric acid. Dilute 125ml of the concentrated acid to 1 litre with water.
Orthophosphoric acid. Sp. gr. 1.75, analytical reagent grade.
Sodium tungstate. Di-hydrate, analytical reagent grade.
Phosphotungstate solution. Dissolve 25g of sodium tungstate in 375ml of water, add 125ml of orthophosphoric acid and mix.
Iso-amyl alcohol. Analytical reagent grade.
Ammonium metavanadate. Analytical reagent grade.
Standard vanadium solutions. 100μg of vanadium per ml : dissolve 115mg of ammonium metavanadate in water, add 250ml of 4M nitric acid and dilute to 500ml with water.

10μg of vanadium per ml : dilute 10ml of the 100μg per ml solution to 100ml with 2M nitric acid.

The decomposition of the sample by fusion with sodium bisulphate is similar to that with potassium bisulphate, but the potassium compound is avoided because, in its presence, a copious, white precipitate of potassium phosphotungstate occurs during cooling at Stage 10 if an aliquot greater than 0.4ml is used. Oxidation takes place atomospherically, but it is not always complete and the sample aliquot must be boiled with concentrated nitric acid before the phosphotungstate solution is added. The optimum concentration range of nitric acid in the final aqueous phase is 2.0 to 3.6M.

When samples are so rich in organic matter that decomposition by fusion is inadequate, digestion with 4ml of concentrated nitric acid and 1ml of perchloric acid (72% w/w, analytical reagent grade) may be applied to a sample weight of 0.25g. After heating to fumes of perchloric acid, the solution is diluted to 10ml with 2M nitric acid. Alternatively, the samples may be ignited prior to fusion.

There is interference from chromium in concentrations greater than 4000 p.p.m., and from molybdenum greater than 2000 p.p.m. Although iron does not give a colour, it retards the development of the phosphotungstovanadate complex and it is, therefore, essential that heating in the water-bath at Stage 9 is continued for a minimum period of 10 minutes.

The standard series ranges from a colourless zero through increasing intensities of yellow colour in the solvent phase, and it is stable for 24 hours. The range covered is 4–1920 p.p.m. when dilution of the solvent phase is carried out, and it may be extended further by use of a smaller aliquot. A productivity of 100 samples per man-day is easily achieved.

This procedure is described by Stanton and Hardwick (1971); an alternative method is given by Ward, Lakin, Canney, et al. (1963).

4. Cold Extraction Methods of Analysis

In a geochemical survey, it is often necessary to determine the amount of an element that is readily available rather than, or in addition to, its total content in the sample. Such a determination might, for instance, distinguish between a metal present as sulphate and as silicate. This is of particular use in sediment surveys. The sample is subjected to a partial attack, often by means of cold reagents, and the results thus obtained are indicative of the mode of occurrence of the element in question. When using colorimetric methods involving solvent extraction in the presence of the solid sample, it is essential to adopt a mixed colour reaction; adsorption of the coloured complex may occur in the presence of clay fractions and/or organic matter, and the actual shade of colour is then matched with a standard series regardless of its lesser intensity. Sometimes adsorption is complete and the solvent phase is colourless, or organic material may impart a strong colour. In either case it is impossible to obtain a result. Dithizone lends itself readily to such analysis, and the methods most commonly used exploit this reagent.

The solvent extraction methods are subject to considerable trouble due to emulsification of clay colloids and organic material, and it may be impossible to observe the colour of the solvent phase. This trouble may be alleviated by the dropwise addition of n- or iso-amyl alcohol, absolute ethyl alcohol or acetone.

For dithizone methods, where extractive titration is employed, if the test-tube is full and the solvent phase remains pink, the contents of the tube are transferred to a stoppered, graduated, 100ml cylinder and the test-tube is cleaned by shaking it with the next addition of dithizone solution. The titration is then continued using 5ml increments of dithizone solution.

Other methods utilizing a partial attack have involved the digestion of the sample with various dilute acids, either hot or cold. The extracts thus obtained are analysed either by colorimetry or by atomic-absorption spectrophotometry. An example of this is the determination of nickel described in this chapter.

The results obtained by these procedures are usualy relative and may not be comparable with those obtained by other procedures. The use of different weights of sample will not necessarily give the same result, and one procedure must be adhered to rigidly throughout a survey. For the purpose of the statistical treatment, the geochemist prefers results to be reported to two significant figures, especially below 10 p.p.m.

Arsenic

Procedure

1. Weigh 0.25g of sample into a test-tube (19 x 150mm) calibrated at 2 and 10ml.
2. Add 2ml of potassium iodide solution.
3. Dilute to 10ml with 0.75% stannous chloride solution.
4. Add 2-4g of zinc pellets and connect a Gutzeit tube to the test-tube without delay, having previously placed a mercuric chloride paper in the head of the arsenic apparatus.
5. Leave for at least 30 minutes, then remove the mercuric chloride paper and immediately compare the confined spot with artificial standards.
6. Arsenic in p.p.m. = 4 x μg of matching standards.

Standards

1. Into test-tubes (19 x 150mm) containing 2ml of potassium iodide solution and calibrated at 2 and 10ml, pipette four aliquots each of 0, 0.1, 0.2, 0.3, 0.4, 0.6, 0.8 and 1.0ml of standard solution containing 5μg of arsenic per ml.
2. Treat as described in Stages 3 and 4 of the Procedure.
3. After 30 minutes, remove the mercuric chloride papers and select the average spot for each quantity of arsenic. Prepare artificial standards by matching these spots with painted filter paper.

Reagents

Hydrochloric acid. Sp. gr. 1.18, analytical reagent grade.

Potassium iodide solution. Dissolve 10g of potassium iodide in 400ml of water.

Stannous chloride. Di-hydrate, analytical reagent grade.

10% stannous chloride solution. Dissolve 25g of stannous chloride in 250ml of concentrated hydrochloric acid.

0.75% stannous chloride solution. Mix 75ml of the 10% solution with 375ml of hydrochloric acid and 550ml of water. Do not keep this solution for longer than two days.

Ethyl alcohol. Absolute.

Mercuric chloride solution. Dissolve 25g of mercuric chloride in 100ml of ethyl alcohol.

Filter paper. Whatman No. 40, or comparable grade.

Mercuric chloride papers. Allow about 15 filter papers to soak for 1 hour in mercuric chloride solution, leave in the dark to dry, then cut into 1cm squares, avoiding the edge of each circle. Store in an air-tight and light-proof container. Prepare freshly each week.

Acetic acid. Glacial, analytical reagent grade.

Lead acetate. Tri-hydrate, analytical reagent grade.

Lead acetate solution. Dissolve 15g of lead acetate in 100ml of water containing 1ml of acetic acid.

Glass wool. Saturate glass wool with lead acetate solution and leave to dry. Pack a loose plug of this into the tube of each arsenic apparatus, and replenish when it is discoloured along half of its length.

Zinc. Pellets, activated with copper, analytical reagent grade.

Paints. Winsor and Newton's water colours: lemon yellow, aurora yellow, and yellow ochre, or similar shades.

Sodium arsenate. $Na_2HAsO_4 7H_2O$, analytical reagent grade.

Standard arsenic solutions. 100μg of arsenic per ml : dissolve 42mg of sodium arsenate in water, add 0.1ml of hydrochloric acid and dilute to 100ml with water.

5μg of arsenic per ml : dilute 5ml of the 100μg per ml solution to 100ml with water when it is necessary to prepare or check the standards.

The arsenic determined by this method is that which is dissolved by an acid solution just less than 4N, at ambient temperature slightly elevated by the reaction of zinc with hydrochloric acid. Sulphide interferes, and usually necessitates frequent replenishment of the glass wool in the apparatus; it may even be of such a high concentration that it is not possible to obtain an arsenic value. The range covered is 0.4 - 20 p.p.m. and the productivity is 200 samples per man-day. The method is based on that of James (1957).

Copper

Procedure

1. Weigh 0.2g of sample into a test-tube (18 x 180mm) calibrated at 5, 7, 9, 11, etc. to 21, 23ml.
2. Add 5ml of buffer solution.
3. Add 2ml of 0.001% dithizone solution.
4. Cork the tube and shake it vigorously for 2 minutes.
5. If the organic phase is pink, add another 2ml of dithizone solution, cork the tube and again shake it for 2 minutes.
6. Repeat the additions of dithizone solution followed by 2 minutes shaking until the colour of the organic phase is between the 3.0 and 3.5μg standards.

7. Compare with a standard series.
8. Copper in p.p.m. = ml of solvent phase x μg of matching standard.

Standards

1. To eleven test-tubes (18 x 180mm) calibrated at 5 and 10ml, and containing 5ml of buffer solution, add, respectively, 0.5, 1.0, 1.5, 2.0, 2.5, 3.0, 3.5, 4.0, 4.5 and 5.0 μg of copper.
2. Add 5ml of 0.001% dithizone solution.
3. Cork the tubes and shake them vigorously for 2 minutes. Store in the dark when not in use.

Reagents

Potassium dichromate. Analytical reagent grade.

Potassium dichromate solution. Dissolve 50g of potassium dichromate in 1 litre of water.

Benzene, Crystallizable grade. Purify by shaking 1 litre with 10ml of potassium dichromate solution for 2 or 3 minutes. Repeat this treatment if the dichromate is obviously reduced. Discard the aqueous phase, wash the benzene four times by shaking it with successive 500ml volumes of water, and then filter through a Whatman No. 1PS paper.

Carbon tetrachloride.

Dithizone. Analytical reagent grade.

0.01% dithizone solution in carbon tetrachloride. Dissolve 40mg of dithizone in about 10ml of acetone, dilute to 400ml with carbon tetrachloride and store in a vacuum flask.

0.01% dithizone solution in benzene. Dissolve 40mg of dithizone in 400ml of benzene and store in a vacuum flask.

0.001% dithizone solution. Mix 40ml of the 0.01% solution in benzene with 360ml of benzene and store in a vacuum flask. Prepare freshly each day.

Tri-ammonium citrate.

Hydroxylamine hydrochloride.

Buffer solution. Dissolve 100g each of tri-ammonium citrate and hydroxylamine hydrochloride in about 500ml of water. Extract with 0.01% dithizone solution in carbon tetrachloride until free from copper, and then extract the excess of dithizone with carbon tetrachloride. Dilute the aqueous phase to 1 litre with water.

Hydrochloric acid. Sp. gr, 1.18, analytical reagent grade

0.5M hydrochloric acid. Mix 40ml of the concentrated acid with 840ml of water.

Cupric sulphate. Penta-hydrate, analytical reagent grade.

Standard copper solutions. 100μg of copper per ml : dissolve 200mg of cupric sulphate in 0.5M hydrochloric acid and dilute to 500ml with this acid.

5μg of copper per ml : dilute 5ml of the 100μg per ml solution to 100ml with 0.5M hydrochloric acid.

The buffer solution has a pH value of 4.5 ± 0.15, and the optimum pH range for the extraction of copper dithizonate with benzene is 3.0 – 4.65 in the final aqueous phase. Alkaline samples may elevate the pH value and lead to interference from other dithizone-reacting metals, particularly zinc, lead, cobalt and nickel. When the pH value exceeds 4.65, interfering elements may arise not only from the sample but also from the buffer solution, especially in the case of zinc. At a pH value of 4.65, 100μg of zinc will give a reaction equivalent to 1μg of copper, whereas the same quantity of cobalt, lead or nickel is undetectable. If the validity of the copper result is in doubt, take a volume of the final organic solvent phase and extract it with an equal volume of 0.02M hydrochloric acid, when lead and zinc dithizonates, if present, will be decomposed. The resulting colour can then be matched as copper, although cobalt and nickel contamination may still exist; however, they are less common contaminants.

In the event of very high copper concentrations, a brown organic phase may be obtained. This results from formation of the secondary dithizonate of copper due to a deficiency of dithizone, in place of the pink, primary compound.

The standard series ranges from the green of unreacted dithizone in the zero standard, through blue between the 3.0 and 3.5μg standards to pink at 5.0μg. It must be prepared freshly each day and stored in the dark when not in use. The concentration range covered is 0.4 - 63 p.p.m. when using a test-tube, and it may be extended by use of a stoppered, graduated cylinder. The productivity is 200 samples per man-day.

This method is based on the procedure described by Holman (1956-7).

Heavy Metals

Procedure

1. Weigh 0.1g of sample into a test-tube (18 x 180mm) calibrated at 5, 6, 7, 9, 11, etc. to 21, 23ml.
2. Add 5ml of buffer solution.
3. Add 1ml of 0.0015% dithizone solution.
4. Cork the tube and shake it vigorously for 15s.
5. If the organic phase is pink, add 1ml of dithizone solution, cork the tube and again shake for 15s.
6. Repeat with 2ml additions of dithizone solution followed by 15s. shaking until the colour of the organic phase is between the 2.0 and 2.5μg standards.
7. Compare with a standard series.

8. Heavy metals (as zinc equivalents) in p.p.m. =
   2 x ml of solvent phase x μg of matching standard.

Standards

1. To twelve test-tubes (18 x 180mm) calibrated at 5 and 10ml and containing 5ml of buffer solution, add, respectively, 0, 0.5, 1.0, 1.5, 2.0, 2.5, 3.0, 3.5, 4.0, 4.5, and 5.0μg of zinc.
2. Add 5ml of 0.0015% dithizone solution.
3. Cork the tubes and shake them vigourously for 15s. Store in the dark when not in use.

Reagents

Potassium dichromate. Analytical reagent grade.
Potassium dichromate solution. Dissolve 50g of potassium dichromate in 1 litre of water.
Benzene. Crystallizable grade. Purify by shaking 1 litre with 10ml of potassium dichromate solution for 2 or 3 minutes. Repeat this treatment if the dichromate is obviously reduced. Discard the aqueous phase, wash the benzene four times by shaking it with successive 500ml volumes of water, and then filter through a Whatman No. 1PS paper.
Carbon tetrachloride.
Dithizone. Analytical reagent grade.
0.01% dithizone solution. Dissolve 40gm of dithizone in about 10ml of acetone, dilute to 400ml with carbon tetrachloride and store in a vacuum flask.
0.015% dithizone solution. Dissolve 60mg of dithizone in 400ml of benzene and store in a vacuum flask.
0.0015% dithizone solution. Mix 40ml of the 0.015% solution with 360ml of benzene and store in a vacuum flask. Prepare freshly each day.
Citric acid. Mono-hydrate.
Hydroxylamine hydrochloride.
Thymol blue. Sodium salt.
Thymol blue solution. Dissolve 40mg of thymol blue in 100ml of water.
Ammonia solution. Sp. gr. 0.88, analytical reagent grade.
Chloroform. Analytical reagent grade.
Buffer solution. Dissolve 45g of citric acid and 10g of hydroxylamine hydrochloride in about 900ml of water, add 5ml of thymol blue solution followed by ammonia solution with mixing until a distinct blue colour is obtained. Dilute to 1 litre with water when the pH value of the solution should be not less than 8.9. Extract with 0.01% dithizone solution until free from dithizone-reacting metals, and remove the excess of dithizone by extraction with chloroform.
Zinc sulphate. Hepta-hydrate, analytical reagent grade.

Standard zinc solutions. 100μg of zinc per ml: dissolve 220mg of zinc sulphate in water and dilute to 500ml.

5μg of zinc per ml : dilute 5ml of the 100μg per ml solution to 100ml with water.

A large number of elements reacts with dithizone in citrate solution at pH 8.9 and upwards, the principal ones encountered in geochemical exploration being cobalt, copper, lead, nickel and zinc. When 10μg of each of these elements are treated as described above, the values obtained are equivalent to 6, 4, 2, 0.2 and 10μg of zinc, respectively.

The values obtained for heavy metals may well be at variance with those for copper, lead and zinc as determined individually. This is a result of the different compositions of the buffer solutions used, which are chosen for their suitability in the colorimetric procedure rather than their efficiency in extracting a particular metal.

This method is based on the procedure of Bloom (1955).

Lead

Procedure

1. Weigh 0.2g of sample into a test-tube (18 x 180mm) calibrated at 5, 7, 9, 11, etc. to 21, 23ml.
2. Add 5ml of buffer solution.
3. Add 2ml of 0.001% dithizone solution.
4. Cork the tube and shake it vigorously for 15s.
5. If the organic phase is pink, add another 2ml of dithizone solution, cork the tube and again shake it for 15s.
6. Repeat the additions of dithizone solution followed by shaking for 15s until the colour of the organic phase is between the 4.0 and 4.5μg standards.
7. Compare with a standard series.
8. Lead in p.p.m. = ml of solvent phase x μg of matching standard.

Standards

1. To twelve test-tubes (18 x 180mm) calibrated at 5 and 10ml and containing 5ml of buffer solution, add, respectively, 0, 0.5, 1.0, 1.5, 2.0, 2.5, 3.0, 3.5, 4.0, 4.5, and 5.0 μg of lead.
2. Add 5ml of 0.001% dithizone solution.
3. Cork the tubes and shake them vigorously for 15s. Store in the dark when not in use.

Reagents

Potassium dichromate. Analytical reagent grade.

Potassium dichromate solution. Dissolve 50g of potassium dichromate in 1 litre of water.

Benzene. Crystallizable grade. Purify by shaking 1 litre with 10ml of potassium dichromate solution for 2 or 3 minutes. Repeat this treatment if the dichromate is obviously reduced. Discard the aqueous phase, wash the benzene four times by shaking it with successive 500ml volumes of water, and then filter through a Whatman No. 1PS paper.
Carbon tetrachloride.
Chloroform. Analytical reagent grade.
Dithizone. Analytical reagent grade.
0.01% dithizone solution in carbon tetrachloride. Dissolve 40mg of dithizone in about 10ml of acetone, dilute to 400ml with carbon tetrachloride and store in a vacuum flask.
0.01% dithizone in benzene solution. Dissolve 40mg of dithizone in 400ml of benzene and store in a vacuum flask.
0.001% dithizone solution. Mix 40ml of the 0.01% solution in benzene with 360ml of benzene and store in a vacuum flask. Prepare freshly each day.
Thymol blue. Sodium salt.
Thymol blue solution. Dissolve 40mg of thymol blue in 100ml of water.
Ammonia solution. Sp. gr. 0.88, analytical reagent grade.
Citric acid. Mono-hydrate, analytical reagent grade.
Hydroxylamine hydrochloride. Analytical reagent grade.
Potassium cyanide. Analytical reagent grade.
Potassium cyanide solution. Dissolve 90g of potassium cyanide in 500ml of water and store in a borosilicate glass bottle.
Buffer solution. Dissolve 100g of citric acid and 20g of hydroxylamine hydrochloride in about 800ml of water. Add 5ml of thymol blue solution followed by ammonia solution with mixing until a distinct blue colour is obtained. These reagents will often be free from lead, but if this is not so, extract with 0.01% dithizone solution in carbon tetrachloride until free from dithizone-reacting metals, and remove the excess of dithizone by extraction with chloroform. Dilute to 900ml with water and store in a polythene bottle. Prepare a supply of buffer solution sufficient for one day's work by mixing in the proportions of 100ml of potassium cyanide solution to 900ml of citrate solution, when the pH value should be 9.5 – 9.8. If the potassium cyanide contains an appreciable amount of lead, this final solution must be purified by treatment with dithizone as described above.
Lead nitrate. Analytical reagent grade.
Standard lead solutions. 100μg of lead per ml : dissolve 80mg of lead nitrate in water containing 1ml of glacial acetic acid and dilute to 500ml with water.

5μg of lead per ml : dilute 5ml of the 100μg per ml solution to 100ml with water.

The method utilizes the fact that the complex cyanides of many metals are more stable than their dithizonates. In particular, cobalt, copper, nickel and zinc do not react with dithizone in ammoniacal citrate medium at a pH value greater than 8.5 when cyanide is present, whereas lead dithizonate is readily formed.

The standard series ranges in colour from the green of unreacted dithizone in the zero standard to the blue at the 4.0μg standard and purple at 5.0μg. The concentration range of 0.4 – 90p.p.m. is covered by use of a test-tube, and the productivity is 200 samples per man-day.

For precautions to be taken when using cyanide see p.6.

Nickel

Procedure

1. Weigh 0.2g of sample into a test-tube (16 x 150mm) calibrated at 5ml.
2. Add 5ml of 0.01M hydrochloric acid saturated with bromine.
3. Stopper the tube and shake it vigorously for 1 hour.
4. Leave to settle.
5. Aspirate the clear solution into the flame of an atomic-absorption spectrophotometer, taking care not to disturb the residue.
6. Calibrate the instrument with standard solutions of suitable concentration prepared in 0.01M hydrochloric acid saturated with bromine.

Reagents

Hydrochloric acid. Sp. gr. 1.18, analytical reagent grade.
Bromine. Analytical reagent grade.
0.01M hydrochloric acid saturated with bromine. Mix 1ml of concentrated hydrochloric acid with 1100ml of water and saturate with bromine.

This method is an attempt to determine the nickel present as sulphide. However, nickel sulphate and carbonate could also be taken into solution, and the nickel determined is more appropriately described as the non-silicate nickel.

Other metals present as sulphide are likely to be taken into solution with various degrees of efficiency, and they may also be determined by atomic-absorption spectrophotometry. Such elements include cobalt, copper, iron, lead and zinc. The sulphides are oxidized by bromine, and the dilute acid solution prevents precipitation by hydrolysis, while being too weak to attack silicates in the cold.

The method is based on that of Davies (1972), another procedure being described by Lynch (1971).

Zinc

Procedure

1. Weigh 0.2g of sample into a test-tube (18 x 180mm) calibrated at 5, 7, 9, 11, etc. to 21, 23ml.
2. Add 5ml of buffer solution.
3. Add 2ml of 0.001% dithizone solution.
4. Cork the tube and shake it vigorously for 1 minute.
5. If the solvent phase is pink, add another 2ml of dithizone solution, cork the tube and again shake it for 1 minute.
6. Repeat the additions of dithizone solution followed by 1 minute of shaking until the colour of the solvent phase is close to that of the 1.5μg standard.
7. Compare with a standard series.
8. Zinc in p.p.m. = ml of solvent phase x μg of matching standard.

Standards

1. To eight test-tubes (18 x 180mm) calibrated at 5 and 10ml and containing 5ml of buffer solution, add, respectively, 0, 0.5, 1.0, 1.5, 2.0, 2.5, 3.0 and 3.5 μg of zinc.
2. Add 5ml of 0.001% dithizone solution.
3. Cork the tubes and shake them vigorously for 1 minute. Store in the dark when not in use.

Reagents

Potassium dichromate. Analytical reagent grade.

Potassium dichromate solution. Dissolve 50g of potassium dichromate in 1 litre of water.

Benzene. Crystallizable grade. Purify by shaking 1 litre with 10ml of potassium dichromate solution for 2 or 3 minutes. Repeat this treatment if the dichromate is obviously reduced. Discard the aqueous phase, wash the benzene four times by shaking it with successive 500ml volumes of water, and then filter through a Whatman No. 1PS paper.

Carbon tetrachloride.

Dithizone. Analytical reagent grade.

0.01% dithizone solution in carbon tetrachlrode. Dissolve 40mg of dithizone in about 10ml of acetone and dilute to 400ml with carbon tetrachloride and store in a vacuum flask.

0.01% dithizone solution in benzene. Dissolve 40mg of dithizone in 400ml of benzene and store in a vacuum flask.

0.001% dithizone solution. Mix 40ml of the 0.01% solution in benzene with 360ml of benzene and store in a vacuum flask. Prepare freshly each day.

Sodium fluoride.

Sodium acetate. Tri-hydrate.
Sodium thiosulphate. Penta-hydrate.
Acetic acid. Glacial.
Buffer solution. Dissolve 5g of sodium fluoride in about 1 litre of water, then in this solution dissolve, by warming gently, 500g of sodium acetate and 125g of sodium thiosulphate, then add 15ml of acetic acid. When cold, extract this solution with 0.01% dithizone in carbon tetrachloride until free from zinc, then remove the excess of dithizone by extraction with carbon tetrachloride. Dilute the aqueous phase to 2 litres with water.
Zinc sulphate. Hepta-hydrate. Analytical reagent grade.
Standard zinc solutions. 100μg of zinc per ml : dissolve 220mg of zinc sulphate in water and dilute to 500ml.

5μg of zinc per ml : dilute 5ml of the 100μg per ml solution to 100ml with water.

The acetate buffer ensures that the pH value of the test solution is within the optimum range of 5.5 – 6.0. Iron forms ferric acetate, rapidly precipitating as the basic acetate but not interfering in the solvent extraction. Aluminium can suppress the reaction of zinc with dithizone, but masking with fluoride precludes this possibility. The reactions of cobalt, copper, lead and nickel with dithizone are suppressed by thiosulphate, which also very slightly suppresses the zinc reaction.

The standard series ranges from the green colour of unreacted dithizone in the zero standard, through blue for the 1.5μg standard to pink at 3.5μg. It must be prepared freshly each day and stored in the dark when not in use. The concentration range covered is 0.4 – 27 p.p.m. and the productivity is 200 samples per man-day.

5. Analysis by Atomic-Absorption Spectrophotometry

Trace analysis by atomic-absorption spectrophotometry is now a well established technique and is frequently applied to geochemical analysis. The range of elements that may be analysed by this method overlaps that of colorimetric procedures, and for some elements the sensitivity is inferior to that obtained by colorimetry. Nevertheless, the rapidity of the technique makes it very desirable where large numbers of samples are involved.

Atomic-absorption spectrophotometry consists in measuring the absorption of radiation by atomic vapour, produced from the sample solution, at a wavelength that is characteristic of the element being determined. The sample solution is aspirated into a flame which is irradiated by light from a hollow cathode lamp, the cathode of which is made of, or contains, the element being determined, and emits light of a wavelength characteristic of this element. Light is absorbed by the atoms of this element present in the flame; the degree of absorption is related to the concentration of the element and is measured photometrically. The instrument is calibrated by aspirating standard solutions treated similarly. Once the sample is in solution it is a matter of only a few seconds to determine the concentration of a particular element.

In some cases a releasing agent is required, but it can often be avoided by the appropriate choice of fuel mixture; several common ions such as aluminium, phosphate, silicate and sulphate suppress the absorption of calcium, but the addition of an excess of strontium or lanthanum to both sample and standard solution minimizes this effect. The fuel used is most important and may preclude the necessity for a releasing agent, but information on this subject is usually available from the instrument manual.

There are many text-books on atomic-absorption spectrophotometry (e.g. Robinson, 1966; Elwell and Gidley, 1966) which should be consulted for the theoretical background. There are also many commercially available instruments; Elwell and Gidley describe sixteen, and a large number of new designs has been marketed since their book was published. The variety of instruments is considerable, and since each manufacturer usually provides a detailed manual, no attempt will be made to describe instrumental procedures. The function of this chapter is to draw attention to the need for decomposing the sample in a manner suited to the individual elements. When atomic-absorption analysis was first used in geochemical exploration it was often presumed that all elements, other than silicon, were taken completely into solution; this is not so. Moreover, many more interelement interferences have been discovered in recent years (e.g. Foster, 1971) some of which can be corrected by measuring the background level using a hydrogen lamp. The most suitable methods of preparing the sample solution are given below, but there is no guarantee that all minerals will be completely decomposed by any one technique.

Cadmium, Cobalt, Copper, Iron, Lead, Manganese, Nickel. Tellurium and Zinc

Procedure

1. Heat 0.2g of sample with 5ml of acid mixture in a test-tube (16 x 150mm) suspended in a hot air-bath (see Fig. 5.1) and evaporate to dryness.
2. Leach with 5ml of 6M hydrochloric acid on a sand-tray or in a boiling water-bath.
3. Dilute to 10ml with water, mix and leave to settle.
4. Aspirate the clear solution into the flame of the instrument, taking care not to disturb the residue.
5. Calibrate the instrument with standard solutions of suitable concentration prepared in 3M hydrochloric acid.

Fig. 5.1

Reagents

Nitric acid. Sp. gr. 1.42, analytical reagent grade.
Perchloric acid, 60% w/w, analytical reagent grade.
Acid mixture. Mix 800ml of nitric acid with 200ml of perchloric acid.
Hydrochloric acid. Sp. gr. 1.18, analytical reagent grade.
6M hydrochloric acid. Mix 240ml of the concentrated acid with 200ml of water, or use the constant-boiling mixture.
3M hydrochloric acid. Mix 120ml of the concentrated acid with 320ml of water.

Several analysts have reported spuriously high lead values due to calcium interference, which may be corrected by measurement of the background level with a hydrogen lamp. A more sensitive method for tellurium is given by Davis, Ewers and Fletcher (1969).

Barium and Strontium

Procedure

1. Weigh 0.2g of sample into a nickel crucible (15ml).
2. Mix with 1g of fusion mixture and fuse.
3. Cool, add 5ml of water and heat gently on a hotplate.
4. Wash the solution and residue into a beaker.
5. Repeat until all the melt has been removed from the crucible.
6. Heat the beaker on a hotplate to ensure that all soluble salts have been dissolved.
7. Filter through a Whatman No. 40 filter paper (9cm diam.) and wash the residue and paper 10 times with sodium carbonate solution.
8. Dissolve the precipitate through the paper with 6M hydrochloric acid into a volumetric flask (25ml) containing 5ml of potassium solution.
9. Wash the paper with 6M hydrochloric acid until the volume is 25ml.
10. Mix, and aspirate into a nitrous oxide – acetylene flame.
11. Calibrate the instrument with suitable standard solutions prepared in 5M hydrochloric acid.

Reagents

Potassium nitrate. Analytical reagent grade.
Sodium carbonate. Anhydrous, analytical reagent grade.
Fusion mixture. Mix 200g of sodium carbonate with 100g of potassium nitrate. The mixture must be fine enough to pass through an 80 mesh sieve.
Sodium carbonate solution. Dissolve 50g of sodium carbonate in 1 litre of water.
Hydrochloric acid. Sp. gr. 1.18, analytical reagent grade.
6M hydrochloric acid. Mix 240ml of the concentrated acid with 200ml of water, or use the constant-boiling mixture.

5M hydrochloric acid. Mix 200ml of the concentrated acid with 240ml of water.
Potassium chloride. Analytical reagent grade.
Potassium solution. Dissolve 20g of potassium chloride in 1 litre of water.

The fusion decomposes barium and strontium sulphates in particular, and it is essential to wash all of the sulphate out of the filter paper otherwise some barium and strontium sulphate will precipitate.

Even with use of a nitrous oxide – acetylene flame, the barium is partially ionized, and the presence of potassium ions suppresses this effect.

Beryllium

Procedure

1. Mix 0.5g of sample with 2g of ammonium fluoride in a silica crucible (25ml).
2. Heat gently until bubbling ceases; break up any lumps with a spatula and continue heating until fumes are no longer evolved.
3. When cool, leach by bringing to the boil with 5ml of 6M hydrochloric acid.
4. Transfer to a test-tube (16 x 150mm), dilute with water-washings to 10ml, mix and leave to settle.
5. Aspirate the clear solution into the flame of the instrument.
6. Calibrate the instrument with suitable standard solutions prepared in 3M hydrochloric acid.

Reagents

Ammonium fluoride.
Concentrated hydrochloric acid. Sp. gr. 1.18, analytical reagent grade.
6M hydrochloric acid. Mix 240ml of the concentrated acid with 200ml of water, or use the constant-boiling mixture.
3M hydrochloric acid. Mix 120ml of the concentrated acid with 320ml of water.

Decomposition of the sample takes place as described by Stanton (1966) for the colorimetric determination of beryllium.

Bismuth

Procedure

1. Weigh 0.5g of sample into a test-tube (16 x 150mm).
2. Add 5ml of concentrated nitric acid.
3. Heat to just below boiling on a sand-tray for 1 hour.
4. Dilute to 10ml with water, and leave to settle.
5. Aspirate the clear solution into the flame of the instrument.

6. Calibrate the instrument with suitable standard solutions prepared in 8M nitric acid.

Reagents

Nitric acid. Sp. gr. 1.42, analytical reagent grade.
8M nitric acid. Mix 500ml of the concentrated acid with 500ml of water.

Most bismuth minerals are decomposed by nitric acid. If silver is required on the same sample, digestion with mercuric nitrate in nitric acid may be used. This method is based on that of Ward, Nakagawa, Harms and VanSickle (1969).

Calcium, Caesium, Lithium, Magnesium, Potassium, Rubidium and Sodium

Procedure

1. Heat 0.2g of sample in a teflon or platinum crucible with 2ml of nitric acid, 2ml of sulphuric acid and 1.5ml of hydrofluoric acid and evaporate to dryness.
2. Dissolve the residue in 2ml of 6M hydrochloric acid.
3. Dilute to 25ml with water; mix and leave to settle.
4. Aspirate the clear solution into the flame of the instrument.
5. Calibrate the instrument with standard solutions of suitable concentration prepared in 0.48M hydrochloric acid.

Reagents

Nitric acid. Sp. gr. 1.42, analytical reagent grade.
Sulphuric acid. 98% w/w Sp. gr. 1.84, analytical reagent grade.
Hydrofluoric acid. 40% w/w, analytical reagent grade.
Hydrofluoric acid. Sp. gr. 1.18, analytical reagent grade.
6M hydrochloric acid. Mix 240ml of the concentrated acid with 200ml of water, or use the constant-boiling mixture.

The use of a releasing agent such as lanthanum or strontium may be required. This procedure is based on that of Ward, Nakagawa, Harms and VanSickle (1969).

Gold and Tellurium

Procedure

1. Mix 25g of sample with 100ml of bromine solution and leave overnight.
2. Boil to expel the excess of bromine.
3. Dilute to 250ml with water and leave to settle.

4. Decant 200ml of the clear solution and place in a separating funnel (250ml).
5. Add 10ml of isobutylmethylketone, stopper the funnel and shake it for 2 minutes.
6. Separate the phases, retaining both of them.
7. Add another 10ml of isobutylmethylketone to the aqueous phase and again extract.
8. Discard the aqueous phase.
9. Combine the ketone phases and wash them twice by extracting with successive 25ml volumes of 2M hydrobromic acid.
10. Discard the aqueous phase, and dilute the ketone phase to 25ml with isobutylmethylketone.
11. Aspirate the ketone phase into the flame of the instrument.
12. Calibrate the instrument by using suitable standard solutions that have been freshly extracted by isobutylmethylketone from 2M hydrobromic acid.

Reagents

Hydrobromic acid. Sp. gr. 1.49, analytical reagent grade.

2M hydrobromic acid. Mix 226ml of the concentrated acid with 774ml of water.

Bromine. Analytical reagent grade.

Bromine solution. Mix 20ml of bromine with 1 litre of concentrated hydrobromic acid.

Isobutylmethylketone.

Gold and tellurium are taken into solution by the bromine and hydrobromic acid, and isobutylmethylketone extracts the gold as bromoauric acid and the tellurium as bromide. The ketone phase is washed with acid to remove iron, some of which is extracted with the gold and tellurium. This method is based on that of Ward, Nakagawa, Harms and VanSickle (1969).

Mercury

Procedure

1. Weigh 0.5g of sample into a test-tube (16 x 150mm).
2. Add 5ml of acid mixture and a few anti-bumping granules.
3. Heat in an aluminium block on a hot-plate at just below boiling.
4. Evaporate until fumes of perchloric acid are copiously evolved.
5. When cold, transfer the solution to a Quickfit conical flask (150ml) washing with water.
6. Dilute to 50ml with water.
7. Add 4ml of stannous chloride solution and quickly connect up the apparatus.

8. Measure the absorbance of the vapour phase using a mercury hollow cathode lamp.
9. Calibrate the instrument using standard solutions of mercury treated similarly.

Reagents

Nitric acid. 72% w/w. Sp. gr. 1.42. Analytical reagent grade.
Perchloric acid. 60% w/w. Analytical reagent grade.
Acid mixture. Mix 800ml of nitric acid with 200ml of perchloric acid.
Stannous chloride solution. Dissolve 50g of the dihydrate in 100ml of hydrochloric acid and dilute to 500ml with water.

Mercury is taken into solution by nitric acid, which together with the perchloric acid oxidizes any organic matter. The test-tube is heated in a block of aluminium which has a hole drilled sufficiently deep to allow the test-tube to stand upright. The heating is carried out in a fume cupboard with a good draught, such that the upper part of the tube is kept cool. This serves to prevent loss of mercury by evaporation. All the nitric acid must be expelled from the sample solution, otherwise the subsequent reduction with stannous chloride will be inhibited. If the solution is evaporated to dryness, mercury will be lost.

The mercuric ions in the sample solution are reduced to the metallic state by addition of stannous chloride, and the elemental mercury is removed from the solution as vapour by flushing out with air. This vapour phase is passed through a gas flow cell situated above the unlit burner of the instrument and irradiated with a mercury hollow cathode lamp, the absorbance being measured.

Most manufacturers supply a suitable apparatus to use with their own particular instruments. The method is essentially that of Hatch and Ott (1968).

Silver

Procedure

1. Weigh 0.2g of sample into a test-tube (16 x 150mm).
2. Add 5ml of mercuric nitrate solution.
3. Heat to just below boiling on a sand-tray for 1 hour.
4. Dilute to 10ml with water, mix and leave to settle.
5. Aspirate the clear solution into the flame of the instrument.
6. Calibrate the instrument with suitable standard solutions prepared in 8M nitric acid.

Reagents

Nitric acid. Sp. gr. 1.42, analytical reagent grade.
8M nitric acid. Mix 500ml of the concentrated acid with 500ml of water.

Mercuric nitrate solution. Dissolve 0.3g of mercuric nitrate in 1 litre of concentrated nitric acid.

Silver is dissolved by digestion with nitric acid and interference from chloride, which may be present in the reagents and the water as well as in the sample, is avoided in the presence of mercuric ions. This procedure is based on that of Ward, Nakagawa, Harms and VanSickle (1969).

6. Analysis by X-Ray Fluorescence Spectrometry

This technique has not been used extensively in geochemical exploration, although it has been exploited frequently in pure geochemical studies. It is particularly useful for the determination of the precious metals, where concentration levels are very low and a high degree of precision and of accuracy are desirable. The elements to be determined are separated from the sample and concentrated into a small precipitate. This precipitate is bombarded with X-rays, and radiation is emitted by the specimen at angles which depend on the element present. The intensity of this emitted radiation is measured at an angle that is characteristic of the element being determined, and the instrument is calibrated by treating known amounts of that element in a manner similar to that adopted for the sample.

The theoretical background to X-ray fluorescence spectrometry is given by Liebhafsky et al. (1960) and Adler (1966). The application to trace analysis is described by Luke (1968), who gives procedures for separating a large number of elements. However, these procedures are not all readily applicable to geological materials, and the most suitable methods are given below.

The essential instrumental conditions are given in Table 6.1.

Table 6.1 Instrumental conditions for X-ray spectrometry

Element	$2\theta^\circ$	Crystal	Counter	Target
As	48.88	LiF 220	Scintillation	Mo
Bi	47.36	LiF 220	Scintillation	Mo
Au	53.40	LiF 220	Scintillation	Cr
Pd	60.02	PET	Flow	Cr
Pt	55.05	LiF 220	Scintillation	Cr
Se	45.80	LiF 220	Scintillation	Cr
Te	109.49	LiF 200	Flow	Cr

All at 50 kV and 48 mA, with fine collimator

Arsenic, Selenium and Tellurium

Procedure

1. Weigh 1g of sample into a beaker (50ml).
2. Add 20ml of acid mixture.
3. Evaporate to fumes of perchloric acid until the volume of the solution is between 1 and 2ml.
4. When cold, add 10ml of 6M hydrochloric acid and heat to boiling.
5. Filter the hot solution through a Whatman No. 540 paper (9cm diam.) into a second beaker (50ml).
6. Wash the first beaker and residue with 6M hydrochloric acid.
7. Add 1ml of arsenic or tellurium solution and mix.
8. Add 5ml of hypophosphorous acid and heat gently for 10 minutes without boiling.
9. Immediately filter through a Millipore filter and wash with 6M hydrochloric acid and then with water.
10. When dry count as appropriate.
11. Interpolate the result from a calibration curve prepared from suitable standard solutions treated as described in Stages 7 to 10 of the procedure.

Reagents

Perchloric acid. 60% w/w, analytical reagent grade.
Nitric acid. Sp. gr. 1.42, analytical reagent grade.
Acid mixture. Mix 800ml of nitric acid with 200ml of perchloric acid.
Hydrochloric acid. Sp. gr. 1.18, analytical reagent grade.
6M hydrochloric acid. Mix 240ml of the concentrated acid with 200ml of water.
Sodium hydroxide. Pellets, analytical reagent grade.
Hypophosphorous acid. 50% w/w.
Arsenic solution. Dissolve 260mg of arsenic trioxide and 2g of sodium hydroxide in about 50ml of water and dilute to 100ml with water.
Tellurium solution. Dissolve 125mg of tellurium dioxide in 200ml of 6M hydrochloric acid.
Millipore filter. 47mm diam., 0.45μ, HAWP 00.

This separation procedure is similar to that described for the colorimetric determination of selenium (Stanton 1966) except that the co-precipitant is arsenic when the tellurium content is required, and tellurium for the arsenic determination. Flocculation of the precipitate must be avoided. For calibration, the smallest amount of each element used should be 5μg.

Bismuth

Procedure

1. Weigh 0.5g of sample into a test-tube (19 x 150mm).
2. Add 1.5g of potassium bisulphate, mix, and fuse until a quiescent melt is obtained.
3. Leach with 5ml of 4M nitric acid on a sand-tray or in a boiling water-bath.
4. Filter through a Whatman No. 540 paper (9cm diam.) into a beaker (50ml).
5. Wash the test-tube and residue with 4M nitric acid.
6. Add 10ml of EDTA solution and mix.
7. Add ammonia solution dropwise until a reddish-brown colour persists.
8. Add 4M nitric acid dropwise until the reddish-brown colour just disappears.
9. Add 1ml of ammonia solution and mix.
10. Add 5ml of potassium cyanide solution and mix.
11. Add 1ml of sodium diethyldithiocarbamate solution and mix.
12. Leave to stand for 10 minutes.
13. Filter through a Millipore filter and wash with water.
14. When dry, count as appropriate.
15. Interpolate the bismuth content from a calibration curve prepared from suitable standard solutions as described in Stages 6 to 14 of the Procedure, including 2ml of iron solution with the standard solutions.

Reagents

Potassium bisulphate. Fused, powder.
Nitric acid. Sp. gr. 1.42, analytical reagent grade.
4M nitric acid. Mix 250ml of the concentrated acid with 750ml of water.
Ammonium ferric sulphate. Analytical reagent grade.
Sulphuric acid. Sp. gr. 1.84, analytical reagent grade
Iron solution. Dissolve 8.64g of ammonium ferric sulphate in 100ml of water containing 10ml of concentrated sulphuric acid and dilute to 1 litre with water. This solution will contain 1mg of iron per ml.
EDTA. Ethylenediaminetetra-acetic acid, di-sodium salt.
EDTA solution. Dissolve 50g of EDTA in 1 litre of water
Ammonia solution. Sp. gr. 0.91, analytical reagent grade.
Potassium cyanide solution. Dissolve 25g of potassium cyanide in 500ml of water.
Sodium diethyldithicarbamate solution. Dissolve 1g of sodium diethyldithiocarbamate in 100ml of water.
Millipore filter. 47mm diam., 0.45μ, HAWP 00.

Most of the discussion on the colorimetric determination of bismuth (p. 5) applies here. For calibration, the smallest amount used should be 10μg. This method is described by Stanton (1971).

Gold, Palladium and Platinum

Procedure

1. Weigh 10g of sample into a beaker (400ml).
2. Add 50ml of bromine solution.
3. Leave to stand overnight then boil off the excess of bromine.
4. Dilute to 100ml with water.
5. Filter through a Millipore filter and wash with water.
6. Return the filtrate to the beaker.
7. Add an excess of stannous chloride solution; 25 - 50ml will be required.
8. Add 1ml of tellurium solution to the cold sample solution.
9. Filter through a Millipore filter and wash with 2M hydrobromic acid.
10. Wash with water.
11. When dry, count as appropriate.
12. Interpolate the results from a calibration curve prepared from suitable standard solutions as described in Stages 7 to 10 of the Procedure, using standard solutions prepared in 2M hydrobromic acid.

Reagents

Hydrobromic acid. Sp. gr. 1.49, analytical reagent grade.
2M hydrobromic acid. Mix 226ml of the concentrated acid with 774ml of water.
Bromine. Analytical reagent grade.
Bromine solution. Mix 20ml of bromine with 1 litre of concentrated hydrobromic acid.
Hydrochloric acid. Sp. gr. 1.18, analytical reagent grade.
Stannous chloride solution. Dissolve 500g of the di-hydrate (analytical reagent grade) in concentrated hydrochloric acid and dilute to 1 litre with this acid.
Tellurium solution. Dissolve 625mg of tellurium dioxide in 545ml of concentrated hydrochloric acid and dilute to 1 litre with water.
Millipore filter. 47mm diam., 0.45μ, HAWP 00.

The discussion on the colorimetric determination of palladium plus platinum (p. 11) applies here, up to the precipitation of the elements, except that coagulation of the precipitate must be avoided. The precipitate must be washed free from chloride, which interferes with the measurement of palladium. Tellurium could be determined by the same procedure using gold as the co-precipitant, but the precipitate must be washed completely free from tin which would interfere.

The separation procedure is described by Thompson (1967), Luke (1968) having demonstrated the suitability of the tellurium precipitate for X-ray fluorescence spectrometry.

7. Analysis by Emission Spectrography*

The methods of analysis so far described in this volume offer techniques for the analysis of single or relatively small numbers of elements at one time. The introduction of the 'multi-element' approach in geochemical exploration, however, has resulted in a need for methods which allow the simultaneous determination of large numbers of elements. Spectrographic methods of analysis offer such a facility.

As with colorimetric analysis, the classical methods of spectrography are relatively accurate but slow; rapid methods, when introduced, essentially on photographic instruments, suffered from poor precision and accuracy. However, recent advances in both techniques and instrumentation have improved the quality of spectrographic analysis considerably, and precisions at least comparable with colorimetric analysis are now obtainable for the majority of elements used in geochemical prospecting. This chapter discusses the adaptation and application of these spectrometric methods to the problems of applied geochemistry.

If a rock, soil or sediment sample is heated to a high temperature in an electric arc, the sample is vaporized and many of the component atoms are excited, one or more of the electrons in each such atom being displaced from the normal orbit to one of a higher energy level. Such a situation is unstable, however, and the electrons return to their original orbital positions, emitting energy in the form of light. Each element emits a characteristic heterochromatic spectrum composed of a number of specific wavelengths in both the visible and the ultraviolet region. Since each atom of a given element behaves in the same way under any particular set of conditions, the total amount of light emitted by each element present is proportional to the number of atoms of that element. The intensity of the spectrum emitted by an element is, therefore, a function of the amount of that element present in the sample excited. The lines of the spectrum of an element are not of equal intensity, and a specific wavelength is employed to measure the spectral intensity; for trace geochemical work, this is usually the most intense line in the spectrum.

*Contributed by Dr C.H. James, Leicester University

Spectrographic instruments, therefore, consist of three main components:

(i) an excitation source to generate the spectrum of the sample;
(ii) a dispersing device to break down the light into its component wavelengths;
and (iii) a recording device to measure the intensity of individual wavelengths.

Most instruments used for geochemical purposes employ excitation by a D.C. arc. The sample to be analysed is placed in a cavity in one of a pair of electrodes of graphite or carbon, between which an arc is struck, and the sample is volatilized. In order to control the nature of the burn within the electric arc, a 'spectroscopic buffer' is added to the sample before it is loaded into the electrode. Various buffers are employed, but a mixture of carbon powder with either lithium carbonate or sodium fluoride is the most common. Such a buffer improves the arc in several ways: by acting as a flux and initiating thermochemical changes it enhances volatilization, while its effect in maintaining a relatively constant temperature in the arc is of special importance. Furthermore, if the quantity of buffer employed is large relative to the sample weight, it tends to eliminate extremes of chemical composition. The sample and buffer are homogenized in a small mixing-mill, and the mixture is loaded by pressing the hollow electrode repeatedly into the heaped mixture on a piece of glazed paper, until the cavity is filled.

Different elements volatilize at different times within the arc, a characteristic known as selective volatilization. When multi-element analysis is to be undertaken, therefore, the excitation time must be very carefully controlled. In classical spectrographic methods this problem is overcome by burning the sample 'to completion', but in the rapid methods, a set burning time of short duration is more usual. When a direct-reading spectrograph is employed, the possibility of adopting different intergration times for separate groups of elements offers another solution to this problem.

The light emitted from the excited sample enters the spectrometer through a narrow slit; this produces a beam of light which is then analysed into its component wavelengths either by a prism or by a diffraction grating. Prisms were prevalent in early instruments and are still very common in photographic instruments, but most direct-reading instruments employ a diffraction grating. The main disadvantage of the prism spectrograph is that the dispersion has a logarithmic character, resulting in a tendency for the spread at the low wavelength end of the spectrum to be very large in comparison with that at higher wavelengths, where the spectral lines may be so close that the resolution of individual lines is difficult or even impossible. Furthermore, owing to the transmissive nature of prisms, different prisms made of alternative materials may need to be employed if the whole spectrum is to be studied. Grating instruments suffer from neither of these problems, a linear dispersion being obtained essentially by reflectance. The linear character of grating dispersion also facilitates the identification of any specific wavelength.

Spectra are recorded either photographically or by a photomultiplier. The photographic plate (or film) used in the former case will accommodate 10 to 20 different spectra. However, this method suffers from a number of disadvantages. Owing to the need for development of the plate, the process is relatively slow. Furthermore, extreme care is needed to ensure that both the emulsion on the plate and the technique of development are as alike as possible from one plate to another. Some degree of correction for variations caused by a discrepancy in these features may be obtained by relating the intensity of the spectral line being read to either a background portion of the spectrum in the immediate vicinity of the line, or by proportioning the reading with that obtained for the line of another element added as an internal standard. Such correction techniques require the use of a densitometer to measure the intensity of the individual lines, and this process makes the technique a great deal slower. Accordingly, the estimation of the quantity of each element present is usually achieved by a visual comparison of line intensities with those on standard plates prepared from a series of synthetic samples of known elemental contents.

In order to extend the range, the light obtained by burning both unknown samples and standards may be passed through a step-sector, a rotating device which impedes the beam for different proportions of time in different positions. Using this technique, each spectrum line has bands of differing intensities commonly in such ratios as 1:1/4:1/16. Since both samples and standards are treated in the same manner, direct comparisons can be achieved. This is normally carried out in a comparator which places an enlarged image of the spectrum of the unknown sample next to a similar image of the standard spectrum, and a visual estimate of the quantity of an element present in the sample can be obtained. Using this technique, precisions of the order of 30 to 45% at the 95% confidence level have been reported (Nichol and Henderson-Hamilton, 1965-6), although precisions as poor as ± 100% have been experienced.

One marked advantage of the photographic technique is flexibility. Since every element records its spectrum on the plate, the latter may be read for elements not originally sought, and the plate itself provides a permanent record of the analysis.

Direct-reading instruments offer a more sophisticated but much more expensive approach to the problem. In such instruments the dispersed light obtained from the spectrometer falls on a circular metal band in which slits have been placed to correspond with the wavelengths of the spectral lines of the elements sought. This light then passes via a prism or mirror onto a photomultiplier tube, the output of which is fed into an electronic integrator. Integration proceeds for a pre-determined time, after which the instrument reads out the results for each element in sequence. Read-out is normally by a visual digital voltmeter, with simultaneous print-out on an automatic typewriter, and in some cases also a paper tape punch.

Using such teqhniques, it is necessary to ensure that the spectrum is maintained in a constant position in the instrument; changes in temperature, and, to a lesser extent, atmospheric pressure, may cause a slight but significant change in the optical characteristics of the light path and results in a misalignment of the slits. To overcome this problem, most instruments have some form of monitor system installed which indicates any changes in the position of the spectrum, so that the necessary re-alignment may be made.

Since only specific spectral lines are recorded in direct-reading instruments, local background corrections cannot be made unless a background channel has been incorporated to measure the intensity of light at a position in the spectrum where no elemental line occurs. In some instruments the integration is terminated when this background reading attains a certain level, while in others an electronic device adjusts the readings obtained for each element by a factor calculated by dividing the background reading by a 'norm'. In practice, however, since the amount of background correction varies over the spectral range, many background lines would be required to obtain a genuine correction, and it is now widely felt that this approach is impracticable.

Early methods of direct-reading spectrometry employed standards to prepare a calibration curve for each element. Such a system requires that the validity of the line be checked at regular intervals by burning high and low samples of known composition, and, if necessary, making adjustments to the electronic controls to bring the line to its correct level. The defect of this approach is that the reading for the control samples are subject to error, so that correction may reduce both the accuracy and the precision of the method. A further disadvantage is the time involved in interpolating values from such curves while the technique is also subject to systematic error.

In some laboratories, therefore, an alternative system is adopted. Standards corresponding to the range of elements sought are burned at intervals during the analysis of unknown samples. At the end of each run, a series of values is available for each element from which a specific calibration curve can be constructed if desired. Such curves are very close to quadratic in character (Çelenk, 1972), and the constants of such a curve may be obtained with a computer using a least-squares regression method. In some cases the coefficient of the squared term in the quadratic equation is so low as to indicate virtually a straight line. Once the coefficients of

Table 7.1 Lower limits of determination obtained by a direct-reading spectrometer using corrected data.

Element	p.p.m.	Element	p.p.m.
Ba	2	Mo	6
Be	1	Ni	9
Bi	1	Pb	8
Co	1	Sn	4
Cr	8	Sr	1
Cu	6	Ti	15
Ga	1	V	3
Ge	1	Zn	11
Li	2	Zr	8
Mn	5		

the quadratic equation for each spectral line are known, the values for the unknown samples may be calculated and printed out.

This approach, however, still suffers from the disadvantages of inter-elemental interference, which may be due to one of two causes: an alteration in the character of the burn (true matrix effect), or the imposition of part of a spectrum of one element upon another. In both cases, it can be shown that a quantitative relationship exists between the quantity of the interfering element present and the effect it produces. Furthermore, in virtually all cases the combined effects appear to be algebraically related so that by summing individual interferences the combined effect may be calculated. Thus, once a detailed study of the interference of each element upon the remainder has been undertaken, calculated corrections may be applied to the data. Punched paper tape outputs from an instrument may be fed directly into a computer, which can calculate the necessary correlations and corrections.

Where possible, however, it is advisable to analyse samples of similar composition in groups, so that synthetic standards, with a base composition similar to that of the unknown samples, may be used for comparison. Thus, the amount of correction required is kept to a minimum and more accurate results are obtained.

The limits of detection vary according to the actual technique employed. For most elements, a detection limit of 5 p.p.m. or less is readily obtainable (see Table 7.1). In certain cases the sensitivity is poor, unless a technique to analyse for the specific element is employed, and even then the limit of detection may not be adequate. Repeated comparisons with standard rocks show results which compare favourably with those obtained by classical and other methods. For most trace elements the precision obtainable is better than ± 40% at the 95% confidence level on uncorrected data. After correction to such data, results considerably better than ± 15% have been reported.

Using photographic techniques, some 25 to 40 samples may be analysed for 10 to 20 elements per 8-hour day with a staff of three. Direct-reading instruments have a higher productivity, a staff of two being capable of analysing 50 to 80 samples per 8-hour day for as many elements as the instrument is equipped to analyse – commonly up to 40. Thus, 3200 elemental determinations per 8-hour day are possible. It must be stressed, however, that the latter figures are only possible if a computer is available to handle the data. Without this facility, the production of even uncorrected data may take considerably longer than the actual analysis, and a productivity only marginally better than photographic instruments may be obtained.

Since individual instruments have their own characteristics it is not possible to provide universally applicable methods of analysis and satisfactory techniques must be developed to meet the requirements of any particular laboratory. However, a detailed description of a photographic technique has been given by Nichol and Henderson-Hamilton (1965-6), while a direct reading method is described by Davenport (1970). A description of computer methods of data handling and inter-element correction is given by Çelenk (1972). Details of other spectrographic techniques are to be found throughout geochemical and analytical literature. Excellent reviews of spectrography in pure geochemistry have been written by Ahrens (1950, 1954), Ahrens and Taylor (1961), and Willard, Merritt and Dean (1965).

References

ADLER, I. (1966). *X-ray emission spectrography in geology.* Amsterdam: Elsevier.

AHRENS, L.H. (1950). *Spectrochemical analysis.* Cambridge, Mass.: Addison Wesley.

– (1954). *Quantitative spectrochemical analysis of silicates.* New York and London: Pergamon.

AHRENS, L.H. and TAYLOR, S.R. (1961). *Spectrochemical analysis.* 2nd edn. Cambridge, Mass.: Addison Wesley.

BAKER, W.E. (1965). *Bull. Australas. Inst. Min. Metall.,* No. 214, 125.

BLOOM, H. (1955). *Econ. Geol.,* **50**,553.

BOWDEN, P. (1964). *Analyst,* **89**, 771.

B.R.G.G.M. (1957). *Méthodes d'analyses utilisées par Bureau de Recherches Géologiques, Géophysiques et Minière en prospection microchemique.* Paris: B.R.G.G.M.

ÇELENK, O. (1972). *Application of computer orientated mathematical and statistical techniques to geochemical exploration data.* Leicester Univ.: Ph.D. thesis.

CRAVEN, C.A.U. (1953-4). *Trans. Instn. Min. Metall.,* **63**, 551.

DAVENPORT, T.G. (1970). *Geochemical studies of the bauxite deposits of the McKenzie Region, Guyana.* Leicester Univ.: Ph.D. thesis.

DAVIS, C.E.S. (1972). *Econ. Geol.,* **67**, 1075.

DAVIS, C.E.S., EWERS, W.E. and FLETCHER, A.B. (1969). *Proc. Australas. Inst. Min. Metall.,* No. 232, 67.

ELWELL, W.T. and GIDLEY, J.A.F. (1966). *Atomic-absorption Spectrophotometry.* 2nd edn. London: Pergamon.

FOSTER, J.R. (1971) Canadian Inst. Min., Special Vol. 11, 554.

HATCH, W.R. and OTT, W.L. (1968). *Analyt. Chem.* **40**, 2085

HOLMAN, R.H.C. (1956-7). *Trans. Instn; Min. Metall.,* **66**, 7.

JAMES, C.H. (1957). *Applied geochemical studies in Southern Rhodesia and Great Britain.* Lond. Univ.: Ph.D. thesis.

– (1970). *Trans. Instn. Min. Metall.,* **79**, B88.

LIEBHAFSKY, H.A., PFEIFFER, H.G., WINSLOWE, E.H. and ZEMENY, A.D. (1966). *X-ray absorption and emission in analytical chemistry.* New York: Wiley.

LUKE, C.L. (1968). *Analytica chim. Acta,* **41**, 237.

LYNCH, J.J. (1971). In: BOYLE, R.W. and McGERRIGLE, J.I., Canadian Inst. Min., Special Vol. 11, 313.

MARRANZINO, A.P. and WARD, F.N. (1958). Conf. Analyt. Chem. Appl. Spectrosc., Pittsburgh, Programme abstract.

MARSHAL, N.J. (1964). *Econ. Geol.*, **59**, 142.
NICHOL, I. and HENDERSON-HAMILTON, J.C. (1965-6). *Trans. Instn Min. Metall.* **74**, 953.
NORTH, A.A. (1956). *Analyst*, **81**, 660.
ROYAL INSTITUTE OF CHEMISTRY (1961). *Laboratory handbook of toxic agents.* London: The Royal Institute of Chemistry.
STANTON, R.E. (1966). *Rapid methods of trace analysis for geochemical application.* London: Edward Arnold.
– (1970a). *Proc. Australas. Inst. Min. Metall.*, No. 235, 101.
– (1970b). *Proc. Australas. Inst. Min. Metall.*, No. 236, 59.
– (1971a). *Proc. Australas. Inst. Min. Metall.*, No. 239, 101.
– (1971b). *Proc. Australas. Inst. Min. Metall.*, No. 240, 113.
– (1975) *Lab. Practice*, **24**, 525.
STANTON, R. E. and HARDWICK, A.J. (1967). *Analyst*, **92**, 387.
– (1971). *Proc. Australas. Inst. Min. Metall.*, No. 240, 115,
STANTON, R.E. and McDONALD, A.J. (1961-2). *Trans. Instn Min. Metall.*, **71**, 27
– (1966). *Analyst*, **91**, 775.
STANTON, R.E., MOCKLER, M. and NEWTON, S. (1973). *J. Geochem. Explor.*, **2**, 37.
THOMPSON, C.E. (1967). *Prof. Paper U.S. Geol. Surv.* No. 575-D, D236.
WARD, F.N. (1951). *Circ. U.S. Geol. Surv.*, No. 119.
WARD, F.N., LAKIN, H.W., CANNEY, F.C. et al (1963). *Bull. U.S. Geol. Surv.*, No. 1152.
WARD, F.N., NAKAGAWA, H.M., HARMS, T.F. and VANSICKLE, G.H. (1969). *Bull. U.S. Geol. Surv.*, No. 1289.
WILLARD, H.H., MERRITT, L.L. and DEANS, J.A. (1965). *Instrumental methods of analysis.* 4th edn. New York: Van Nostrand.

Index